Rangeland Desertification

Advances in Vegetation Science 19

Rangeland Desertification

Edited by

OLAFUR ARNALDS

and

STEVE ARCHER

Kluwer Academic Publishers

DORDRECHT / BOSTON / LONDON

Library of Congress Cataloging-in-Publication Data

ISBN 978-90-481-5359-6

Published by Kluwer Academic Publishers,
P.O. Box 17, 3300 AA Dordrecht, The Netherlands.

Sold and distributed in North, Central and South America
by Kluwer Academic Publishers,
101 Philip Drive, Norwell, MA 02061, U.S.A.

In all other countries, sold and distributed
by Kluwer Academic Publishers,
P.O. Box 322, 3300 AH Dordrecht, The Netherlands.

Printed on acid-free paper

Printed in the Netherlands.

Table of contents

Preface vii
Ms. Vigdís Finnbogadóttir, Former President of Iceland
Introduction 1
Olafur Arnalds and Steve Archer

Processes

Desertification: an appeal for a broader perspective 5
Olafur Arnalds
Sress, disturbance and change in rangeland ecosystems 17
Steve Archer and Chris Stokes
Viewing rangelands as landscape systems 39
John A. Ludwig and David J. Tongway
Hydrologic effects on rangeland degradation and restoration processes . . 53
Thomas L. Thurow
Erosion models: use and misuse on rangelands 67
Frederick B. Pierson, Jr.
Desert rangelands, degradation and nutrients 77
Kris M. Havstad, J.E. Herrick and W.H. Schlesinger
Assessing and monitoring desertification with soil indicators 89
David Tongway and Norman Hindley
Scaling up from field measurements to large areas using the Desertification Response Unit and Indicator Approaches 99
Anton Imeson and Erik Cammeraat
Agricultural and ecological perspectives of vegetation dynamics and desertification 115
M. Timm Hoffman

Policy

The United Nations data bases on desertification 131
W. Franklin G. Cardy
The implementation of soil conservation programmes 143
David Sanders
Evolution of rangeland conservation strategies 153
Andres Arnalds
Policy and law for rangeland conservation 165
Ian Hannam
Rangelands issues and trends in developing countries 181
Hamid Narjisse
The United Nations Convention to Combat Desertification: constraints to implementation in Eastern Africa 197
Naftali Manddy Onchere

Preface

Ms. Vigdís Finnbogadóttir

The well-being of the soil and what it nurtures is vital for the future, a grim lesson from the past, a lesson we know so well in Iceland. The fate of nature is a subject that has always been close to my heart. I have, therefore, always been ready to encourage, to help, and to challenge our people, who are devoted to the protection of the soil.

During one of my terms of office, a group of people met at the presidential residence, Bessastaðir, to discuss the fate of Icelandic rangelands, – and the rangelands of the world. And thus, the idea was born to organise an international workshop on rangeland desertification. It is therefore with great pleasure that I write a few words at the onset of this book, which is a direct result of the workshop, and this meeting at Bessastaðir.

Why are there deserts in Iceland? On first impression, Iceland's mid-ocean environment would seem to be an ideal setting for plants to thrive and reproduce, with more than enough rain to water them and secure the soil, and where there is certainly no danger of them withering in the heat.

Maybe, too, the guardian nature spirits that were said to have been here when the first settlers arrived 1100 years ago (and now illustrate our national coat of arms, the dragon, the eagle, the bull, and the giant) could be expected to protect their home from destruction, as they presumably had been doing ever since this island rose from the ocean more than 20 million years ago. But no. The guardian spirits could not overcome the combined destructive powers of man and the hostile environment.

The Icelandic ecology is extremely sensitive, with many natural forces that make its rangelands fragile. The climate is colder than in many other countries, but that does not make deserts, – a cooler climate often helps to preserve the moisture in the soil. Volcanic eruptions are frequent and often cause massive damage, but nature is adapted to heal such wounds to its surface. And the sand from the glaciers, brought down in catastrophic floods when a volcano erupts underneath the ice mass, – yes, this is a destructive force, but certainly not the only one to blame for the crisis that Iceland's vegetation is facing.

Nature is resilient, it can take the blows exerted on it, but only up to a point. And when man, in times of old, not knowing the extreme fragility of the Icelandic rangelands and the volcanic soils, started to cut the shrublands and graze the highlands to make his living from the land, the pressure became too great.

Iceland has deserts. Nearly half of the country is barren wasteland, where nature provides neither food nor shelter from the howling North Atlantic winds. And we know, with certainty, that this has not always been so. Man, in times of hardship, aided by cooler climate and fierce natural forces, has altered the face of the country and its ability to provide for its people.

Figure 1. Ms. Vigdís Finnbogadóttir with participants of the Rangeland Desertification Workshop during a field excursion. Photo G. Johannesson.

There is a remarkable passage in the *Book of Icelanders* written by the historian Ari the Learned, in the early twelfth century, in which he describes the land found by the settlers as being "covered with woods from mountain to shore." In the Sagas, – our renowned medieval stories written in the vernacular, – there are trees everywhere. People would hide there from their enemies for days, all over the country; some of the woods were too dense even to enter. In the classic work of Icelandic literature, *Njal's saga*, Njal himself and the other hero, Gunnar, jointly owned a wood which became the occasion of a fateful feud.

When Gunnar was about to go into outlawry and was riding away, from his farmstead to the shore to sail to Norway, he looked back and suddenly spoke the words that everyone in Iceland knows and admires – "So lovely is the hillside that it has never seemed to me as lovely as now, with its pale fields and mown meadows; and I will ride back home, and not go anywhere at all" – whereupon he turned back to face certain death. Interestingly, many of the Sagas say that such –and – such a place was "covered with woods at that time" which suggests that something had changed radically over the three hundred years between the action of the Sagas and their writing in the thirteenth century. Numerous place-names, where there is now no vegetation cover indicate that they were once green and flourishing: *holt*, wooden hill; *hagi*, hagen, field; and *vellir*, grassy plains.

Figure 2. Ms. Vigdís Finnbogadóttir with Workshop participants in the field. Photo G. Johannesson.

Did man realise the damage he caused? It is stated in an ancient law that "it is forbidden to graze the commons so much as to cause their value for grazing to diminish." Sustainable harvesting, then, is not a new concept. And there is more recent law to the same effect. But somehow, man is slow to learn. Eroded areas and degraded highland deserts are still grazed, even though we, as a prosperous nation, do not need to do this, and even though we have the knowledge to produce enough food, without damaging the land, in areas which are clearly well suited for grazing by livestock.

Today we know better, but we cannot blame our ancestors for the way they used the land. They struggled for 1000 years simply to survive, and food production controlled the number of children the nation could foster. Fortunately, this is no longer the case.

The story of the degradation of fragile land and the desertification of Iceland, for one thousand years until this century, is in many ways similar to what is happening in many places in the world today. The result is often the tragedy that has struck the dry areas of earth in recent times: famine. With hunger, when nature fails to provide, comes war and social unrest, as we are seeing in parts of Africa even today.

The Icelandic story is not only an account of losses and mishaps. The predecessor to the Icelandic Soil Conservation Service was established as early as 1907, and it is therefore one of the oldest operating soil conservation institutes in the world, if not *the* oldest. And there have been many successes: encroaching sand has been restrained, we know how to stop and prevent soil erosion, we know how to manage the land. We are continually learning more about how to establish lush vegetation cover, for multiple use by people and animals.

Iceland's achievements may not be large on a global scale, but they are great all the same. And they have a symbolic value as well as a practical one. If we can change our deserts into green land up here on the edge of the Arctic Circle, at the border of the habitable world, we can send a message to the rest of the world that this is actually possible anywhere. And in fact that message would not only be aimed at the rest of the world, but just as much at the Icelanders themselves, who for centuries did not try to grow anything here because they were convinced that it could not be done.

Icelanders have mostly been focusing on their own problems, but they have gained knowledge that can be shared and put to good use elsewhere. Their work towards understanding degradation of Icelandic ecosystems earned them the Nordic Nature and Environmental Award in 1998. And equally, we can certainly learn more from other countries. International cooperation is vital in the world-wide struggle against desertification.

Environmental scientists perhaps shoulder one of the greatest responsibilities of all people today: to study, to educate, to provide means to heal the wounds that mankind has inflicted on the Earth, to harness knowledge in order to make a better world for us all to live in. I have endless admiration for such scientists because, in the final analysis, they seem to be motivated by exactly the same classical love and reverence for the "pale fields and mown meadows" that have become an intergral part of the Icelandic national identity.

This book is the fruit of a meeting that took place in my office, some years ago. I am pleased that it has turned out to be such an excellent scientific contribution. On behalf of all of us involved in preparing the workshop and this publication I thank all of you that contributed to such high quality work.

Vigdís Finnbogadóttir

Vigdís Finnbogadóttir[1]

---

[1] Vigdís Finnbogadóttir is currently chairman of the UNESCO's World Commission for Ethics in Scientific Knowledge and Technology. She is the recipient of the UN-FAO *Ceres* Award.

Introduction

Olafur Arnalds[1] and Steve Archer[2]
[1] *Agricultural Research Institute, Keldnaholt, IS-112 Reykjavik, Iceland*
Tel: 354 577 1010; Fax: 354 577 1020; E-mail: <ola@rala.is>
[2] *Department of Rangeland Ecology & Management, Texas A&M University, College Station, TX 77843-2126, USA*
Tel: 409 845 7332; Fax: 409 845 6430; E-mail: <sarcher@vms1.tamu.edu>

The ever-increasing demand for food and natural resources by a rapidly growing human population has exerted environmental stress resulting in widespread ecosystem degradation. An extreme form of such degradation, termed 'desertification', is estimated to affect the living conditions of about one billion people. As a result, this topic spawned the United Nations Convention to Combat Desertification (UN-CCD) in 1994.

The term 'desertification' encompasses a variety of processes and is driven both by natural and anthropogenic forces. Desertification has occurred in most regions of the world, cutting across a broad spectrum of contrasts in climate, ecosystem types, land uses and socio/economic settings. The complexity of this phenomenon has challenged our ability to categorize, inventory, monitor and repair the condition of the land. Short-comings in communication and understanding are magnified by the improper, incomplete or 'out of context' transfer of knowledge from one region or land use category to another. One of the most important distinctions to be made in relation to land degradation is between cultivated land used for annual crop production and 'rangelands'. Rangelands represent a variety of ecosystems and landforms not suited for intensive agriculture or forestry, because of limitations imposed by climate, soils or topography (Stoddardt et al., 1975; Holecheck et al., 1989). Grazing by free-ranging livestock is the traditional primary use of the world's rangelands. However, there is growing recognition of the importance of these vast acreages for wildlife habitat, hydrology and ground water recharge, recreation and aesthetics.

Historic approaches to halting, mitigating or reversing rangeland degradation were agronomically-based rather than ecologically-based. Agronomic approaches were typically intensive, costly and non-sustainable. As such, they were ill-suited to extensively managed rangelands characterized by variable or extreme climatic conditions, poor soils, and/or rugged topography. Agronomic efforts at rangeland improvement often consisted of practices such as broadcast seeding and fertilization with little regard for spatial and temporal heterogeneity or the status of underlying ecosystem processes that promote or retard degradation and restoration.

In recognition of these short-comings, a group of about 80 experts from over 40 countries were assembled in Iceland in September 1997 for a workshop on rangeland desertification. The goal of the workshop was to bring together a broad spectrum of scientific expertise representing bioclimatically and culturally diverse regions to compare and contrast ecological perspectives on rangeland desertification.

Why convene a desertification workshop in Iceland? As noted elsewhere in this volume, severe land degradation has radically impacted most of Iceland's rangeland ecosystems. Iceland thus exemplifies the fact that the problem of desertification extends beyond Africa and the dryland regions of the world. Ongoing degradation continues in much of Iceland, yet there have been significant scientific and social advances in combating desertification. Thus, there is both cause for concern and reason for optimism. In this context, Iceland provided an appropriate, and somewhat extraordinary setting for this workshop.

The experts that attended the workshop represented many disciplines and geographical regions and shared a common interest in severe land degradation. During the workshop, principles of land degradation were discussed and illustrated by a diverse array of case studies from all regions of the world (Figure 1).

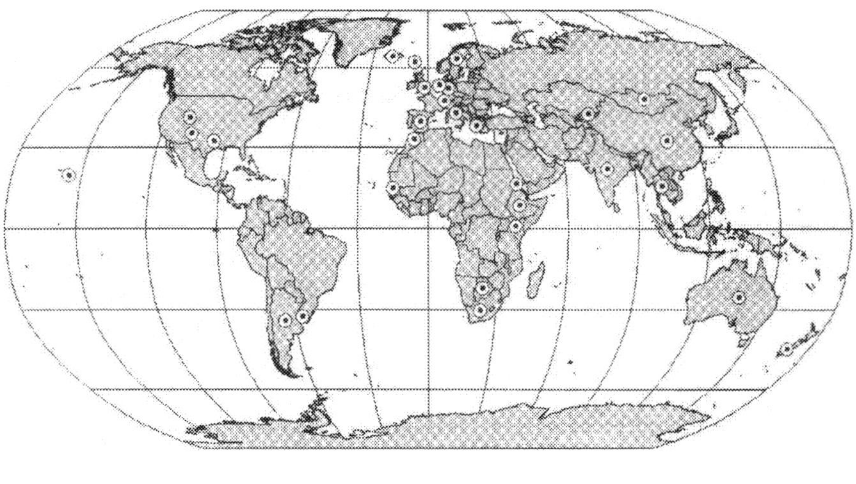

Figure 1. Geographical distribution of degradation research reviewed at the 1997 Rangeland Desertification Workshop in Iceland.

The wealth of information introduced at the workshop could not be presented in a single publication. We, as editors assisted by inputs from an organizational committee, were therefore faced with the task of dividing the papers into two volumes. The volume presented here focuses on concepts and principles. The second volume, published as RALA Report No. 200, summarizes workshop discussions and recommendations and contains a compilation of case studies. As such, it represents a unique documentation of rangeland desertification in many countries and regions of the world.

This volume is divided into two sections. The chapters in the first section explore the spatial and temporal aspects of disturbance interactions, thresholds and non-linear change with respect to vegetation, hydrology, nutrient cycling and erosion. Chapters in the second section of the book are dedicated to socio-economic constraints, remedies and approaches for preventing and reversing degradation. It begins with an overview of United Nations databases on desertification, followed by chapters discussing approaches for implementing conservation practices. A concluding example shows how environmental accountability can be woven into the policy and law of a society. Reversal of the effects of desertification is most difficult in countries with limited resources. One chapter articulates the problems facing developing countries; another describes the constraints to implementing the articles of the UN-CCD in Africa.

ACKNOWLEDGEMENTS

We thank Ms. Vigdís Finnbogadóttir, former President of Iceland for her support of this project from beginning to end. We also thank David Sanders, Denis Peter, Bjorn Sigurbjornsson, Magnus Johannesson, Sveinn Runolfsson, Thorsteinn Tomasson, Halldor Thorgeirsson, Andres Arnalds, Asa L. Aradottir, Ulfur Bjornsson, Einar Gretarsson, Steinunn Geirsdottir, and the numerous Icelanders involved 'behind the scenes' in organizing the logistics of the workshop, field trips and this publication. Both books from the workshop were typeset by Tryggvi Gunnarsson at RALA with the help of Dagny E. Arnalds. We are very thankful for their good work.

The chapters in this book were subject to anonymous peer-reviews. We are grateful to those who generously donated their time for this thankless task.

The workshop was funded by the Icelandic Ministry of Agriculture and the Ministry for the Environment, the Agricultural Research Institute and the Soil Conservation Service. Additional funds were provided by the European Commission. We thank all these organizations for their important contributions.

REFERENCES

Arnalds, O. and Archer, S. (Eds.) 1999. Proceedings of the Rangeland Desertification, International Workshop, September 1997, Iceland. RALA Report No. 200. (In press).
Holechek, J.L., Pieper, R.D. and Herbel, C.H. 1989. Range Management: Principles and Practices. Prentice Hall, Endlewood Cliffs, New Jersey.
Stoddardt, L.A., Box, T.W. and Smith, A.D. 1975. Range Management. McGraw-Hill Book Company, New York, NY.

Figure 2. Two participants of the Rangeland Desertification Workshop observing Icelandic desert. Photo G. Johannesson.

ICELAND

ice
fire
floods
eroded lands
   Iceland
young landscapes
old culture
centuries of Norsemen
carved woodlands
into modern cities
geothermal heat
pure water

sagas
songs
traditions
national self image
of mown meadows
future green pastures
in the minds of
   scientists
      politicians
         poets

Thadis Box

Desertification: an appeal for a broader perspective

Olafur Arnalds
Agricultural Research Institute, Keldnaholt, IS-112 Reykjavik, Iceland
Tel: 354 577 1010; Fax: 354 577 1020; E-mail: <ola@rala.is>

ABSTRACT The term 'desert' has many meanings, but usually refers to an area with a certain climate, vegetation cover, or desolation. 'Desertification' is a vague and often confusing concept because of the many meanings of the term 'desert'. The current definition by the United Nations confines desertification to arid areas. This rather narrow definition limits political and economic actions and constrains programs aimed at combating desertification or reversing land degradation. In this paper, an Icelandic case history is used to illustrate the limitations associated with climatologically-based definitions of desertification. Severe land degradation can lead to the formation of barren land, a desert, in any climate. Desertification is often initiated when ecosystem resilience is reduced through factors associated with drought and/or human activities. However, other factors, such as cold spells, extreme weather events, volcanic eruptions and other environmental stresses can be equally or more important. Severe degradation of ecosystems in Iceland has resulted in the formation of extensive barren deserts in spite of humid climate. The Icelandic example also illustrates that the loss of soil water storage capacity can be as serious a limitation to ecosystem function in humid climates as it is in dry climatic regimes.

It is argued that the climatologically-based definition of desertification used by the UN-Convention to Combat Desertification (UN-CCD) has many negative consequences. Severe land degradation is a global problem not restricted to arid zones. As a result of its narrow definition, the UN-CCD may hamper the development of international, social, political, and scientific programs aimed at combating desertification. Evolution of the CCD from its current regionally limited concept towards a more comprehensive framework which embraces all severe land degradation, is needed. Such an evolution would enhance communication, promote research and help to counter land degradation at the global level.

Key words: CCD, land degradation, desertification, Iceland.

1. INTRODUCTION

Degradation and desertification of the World's land resources affects the livelihood of today's human population and that of the generations to come. Desertification has devastated the productivity and biodiversity of large areas, and the damage is often irreversible when measured on the time-scale of the human lifespan. Desertification may already be causing damage to the world's ecosystems at the same scale as has been predicted for global warming (Dale, 1997). Furthermore, extensive land degradation may be a major factor in the alteration of Earth's climate (e.g., Bolle, 1995). This threat to the World's environment was the subject of a recent United Nations convention, the Convention to Combat Desertification (UN-CCD) (see Cardy, this volume).

The study of desertification is relatively new. The name appeared first in a scientific text about 50 years ago (Auberville, 1949). The affected areas are characterized by large differences in natural conditions, but are normally associated with arid climate. The desertification concept is still evolving. The humid but intensively desertified Icelandic landscapes provide an alternative perspective as compared to arid areas to investigate the concept of desertification.

2. DESERTIFICATION IN ICELAND

Iceland is a 103,000 km^2 island in the North-Atlantic Ocean. The climate, strongly influenced by the warm Gulf Stream, is described as cold temperate in the lowlands and sub-arctic in the highlands. Permafrost is nearly absent. The country is mountainous with lowland areas along the coastline and river plains. The island is humid in most areas. Rainfall generally varies between 600 and 2000 mm yr^{-1} in lowland areas, but large tracts of North-east Iceland receive less than 600 mm. Despite the low evapotranspiration and relatively high rainfall, a large percentage of Iceland's terrestrial ecosystems have been devastated since the arrival of man to the country, about 1100 years ago (Thorarinsson, 1961; Arnalds, 1998). The consequence has been the formation of landscapes which are almost totally barren; or deserts. These barren areas do not fit the climatic definition of "polar deserts" (e.g., Rieger, 1983), as the climate is much milder and more humid than found in polar areas.

Icelandic ecosystems evolved in the absence of large grazing animals. Fully vegetated ecosystems covered most of the country when man arrived and initiated livestock grazing and wood harvesting. Sources for reconstructing past ecosystem structures include pollen analyses, historical records, soil remnants, and relic vegetation (e.g., Einarsson, 1963; Thorarinsson, 1961, 1981; Arnalds, 1987; Hallsdottir, 1995; Kristinsson, 1995; Gisladottir, 1998). These reveal that a large portion (perhaps >25%) of the Icelandic lowlands and lower highlands were once covered with woodlands of birch and willows, and that most of central Iceland was mantled with fertile Andosols and vegetation.

Severe desertification appears to have began soon after the settlement about 1125 years ago (Thorarinsson, 1961, 1981; Gudbergsson, 1975). The main cause for the massive ecosystem degradation is believed to have been animal grazing and wood harvesting. The soils, mostly Andosols, were very susceptible to erosion by wind and water, and to cryogenic processes (Arnalds, 1990; Arnalds et al., 1995; Figs. 3 and 4 in Archer and Stokes, this volume). The surface was subjected to frequent volcanic ash deposition (Thorarinsson, 1961; Sigbjarnarson, 1969; Magnusson, 1994) which intensified eolian processes where the vegetation had been disturbed by utilization. The climate was already becoming cooler when man arrived (Bergthorsson, 1969) and this cooling trend continued long after the arrival of man. Sand encroachment on vegetated land also played a major role, especially in the highlands (Arnalds et al., 1997). The cumulative effect of cooler climate and increased eolian deposition added to the susceptibility of Icelandic soils to erosion (Arnalds, 1999).

The desertified soil surface has quite different soil properties from the former Andosol cover. It lacks nutrients, has very limited water holding capacity, and maintains <5% vegetation cover (Arnalds et al., 1987). More than 37,000 km^2 of the 103,000 km^2 land area are now classified as barren deserts, with additional 15,000 km^2 having limited plant production (LMI, 1993). Birch woodlands now cover only about 1% of the country (Sigurdsson, 1977). Most of the barren desert surfaces, which characterize present day landscapes, have formed during the past 1100 years.

The Icelandic language has two special terms for deserts. The first is 'auðn' (plural 'auðnir'), which is related to 'auður', meaning "empty" or "deserted". The second term, 'eyðimörk', can be translated as "deserted area" and is commonly used to describe the barren landscapes of the world's arid regions. The term 'auðn' is more often applied to the barren Icelandic landscapes. The Icelandic 'auðn' has traditionally been used synonymously with "desert" in English literature (i.e., Anderson and Falk, 1935).

The results of desertification in Iceland have been similar to those reported world-wide: severely reduced productivity; increased demands for food from the land with population growth; and starvation. Social unrest and loss of independence to Norway in 1262 AD (and later to Denmark) is often attributed to the devastation of Icelandic ecosystems (e.g., Sigbjarnarson, 1969). Desertification continued through the middle-ages and is presently recognized as a major environmental problem. Iceland regained independence from Denmark in 1944. The Icelandic Soil Conservation Service was established under Danish rule as early as 1907, to battle encroaching sand as its first aim (A. Arnalds, this volume).

3. WHAT CONSTITUTES A 'DESERT' AND 'DESERTIFICATION'?

A mutual understanding of the desertification concept is vital for communication among scientists, administrators, politicians, and the general public. It has been said that "*no meaningful assessment can be carried out without a clear-cut definition of the problem, hence the need for an unequivocal voice*" (UNEP-DC/PAC, 1990). Several authors and organizations have stressed the need for a clear definition of desertification in order to advance scientific understanding and to promote international co-operation and remedies (e.g., UNEP DC/PAC, 1990; Behnke and Scoones, 1993; Rubio, 1995). Binns (1990) emphasized the need to investigate theoritical definitions of "desertification" and several comprehensive texts have been written about the nature and evolution of the term (Mainguet, 1994; Odingo, 1990; Rubio, 1995; Thomas and Middleton, 1994).

There have been many attempts to weave contrasting perceptions of desertification into a single definition. Even so, no definition has gained full acceptance and the desertification terminology is still subjected to debate. Thornes (1995) noted that most texts on desertification begin with a discussion of definitions.

3.1 Desert as a dry area

To many, the term "desert" simply means an arid area, vegetated or not, with set climatic boundaries. Annual precipitation is often used to define deserts as is done in the Harper Encyclopedia of Science (Newman, 1967), the Dictionary of Geology (Whitten and Brooks, 1974) and the Oxford Dictionary of Natural History (Allaby, 1985). Desertification in this sense would simply imply a decline in rainfall, i.e., to become more arid. Similar are definitions based on indices of aridity, as represented by the ratios of precipitation, evaporation and sometimes transpiration (e.g., Cooke et al., 1993).

The vast majority of deserts will, by people's perception, always be in the arid regions of the world, may they be vegetated (as are many US deserts) or barren seas of sand. When it comes to desertification, however, it should be kept in mind that many dryland ecosystems are deserts based on a climatic definition, before degradation alters the ecosystems. Relating desert and desertification by some measure of aridity can therefore be difficult or impossible.

Set climatic boundaries provide the basis for the UN Convention to Combat Desertification in another way (CCD, 1994; see also Rubio, 1995): "*Desertification is land degradation in arid, semi-arid and dry-subhumid areas resulting from various factors, including climatic variations and human activities.*" This definition emphasizes land degradation, but within certain climatic boundaries. Many have argued strongly for such limitations, e.g. "*dry ecosystems turned into desert*" (Mainguet, 1994).

3.2 Desert as an area of limited plant growth or production

The term 'desert' has often been defined on the basis of vegetation cover or productivity (low production or the lack of plant cover). Definitions of desert and semi-desert biome types are based on structure or physiognomy which is a response to environmental features (Whittaker, 1975). Desertification could then be defined as long-term reduction in vegetation cover or productivity. Changes in ecosystem function can also serve as the basis for the definition of desertification. Glantz and Orlovsky (1983) concluded that "*with all factors cited in the existing definitions, desertification would encompass most kinds of environmental changes related to productivity*".

Definitions of desertification such as "*spread of a desert*", "*intensifying the desert*", or "*dry ecosystems turned into desert*", are often in reference to reductions in vegetation cover. Many of these systems "*turned into desert*" would have been classified as deserts before any ecosystem changes occurred, based on climatic factors. While decreased vegetation cover is often among key attributes of desertification, it is important to note that many climatically defined deserts have considerable vegetation cover while others are barren. This makes definitions of desertification based on changes in vegetation cover often difficult to apply.

'True deserts' and 'natural deserts' are terms that have been used in contrast to an induced 'desert condition'. The emphasis on 'desert condition' is reflected in the first UN definition of desertification: "*...can lead ultimately to desert-like conditions*" (UN, 1977). Glantz and Orlovsky (1983) pointed out that 'desert conditions' cannot be created in a desert, only at its fringes. Thomas and Middleton (1994) stated that the new and improved UN definition (CCD, 1994) "*....firmly returns to the desert margins*". The perception that desertification mainly occurs at the desert fringes is debatable and has contributed to the infamous 'marching desert debate' (see Forse, 1989; Binns, 1990; Helldén, 1991; Mainguet, 1994; Pearce, 1992; Thomas and Middleton, 1994).

3.3 Desert as a desolate area

The word 'desert', the root of 'desertification', has a Latin origin and describes a desolate or deserted condition. The Latin word has Egyptian roots connotating abandonment (see Mainguet, 1994). This meaning, a desolate or sparsely populated area, was perhaps the original intention of Auberville (1949) when he coined the term desertification.

It can be argued that moving away from rainfall limitations to a more inclusive definition, as is reflected by the word 'desolate', would generate several conceptual problems. Many mountain, forest steppe, and tundra regions, often in near pristine condition, are sparsely populated and are therefore desolate. It is undoubtedly unacceptable to most people to describe these areas as desert areas, especially those that are covered with vegetation. Further, it would provide a poor basis for the inventory of desertified areas. The positive aspects of such a definition of desertification lies in avoiding the problems associated with arioity-based definitions discussed earlier.

No single, conclusive ecological definition of the term 'desert' has been accepted. The different perceptions of the term "desert" can be viewed as contrasting paradigms that complicate the discussion about desertification. More than one perception may be used simultaneously without making a distinction between different and even contrasting meanings. The "desert" is at the root of the "desertification" concept and this makes it difficult to define the term "desertification". The pursuit of global definition of desertification may ever be in appropriate and has been questioned (e.g., Perez-Trejo, 1992). The Icelandic example suggests that the definition of the term 'desert' should not *solely* be based on aridity, but should rather include barren areas wherever they occur. This would make the term "desertification" more robust, in that the process can occur in any climatic zone.

4. GLOBAL RECOGNITION OF SEVERE LAND DEGRADATION

4.1 Limitations to the UN definition of desertification

The definition provided by the UN-CCD on desertification is very important, as it defines the scope of the Convention. During the negotiations for the Convention, it was decided to limit the focus on dryland areas of the earth, i.e., "... *in arid, semi-arid and dry-subhumid areas*". An important reason for including climatic limitation in the UN definition of desertification was that the UN Convention to Combat Desertification was primarily meant to focus on the African situation in a pressing search for relief. This is reflected in the subtitle of the convention: "*in those countries experiencing serious drought and/or desertification, particularly in Africa*". In other words, global politics have shaped the definition. However, as discussed earlier, climatic restrictions of the term desertification are among the key reasons for the continuing debates about definitions.

Severe land degradation leading to nearly total devastation of ecosystems does occur outside the regions specified in the UN definition, as exemplified by the Icelandic case history. Severe degradation is also common on slopes in humid areas subjected to intense land use.

The climatic constraints of the UN-CCD definition have other limitations. Arid and semi-arid ecosystems are adapted to variations in annual rainfall. Barren areas may form during drought periods, but recover when rains resume. This was well expressed following the Sahelian droughts 1968–1974 and 1979–1984. There is ample evidence that the productivity of these ecosystems was restored with the return of rain (e.g. Helldén, 1991), as long as ecological disturbance thresholds associated with human activities were not exceeded (see Archer and Stokes, this volume).

4.2 Arid areas or marginal lands

The central concept of the UN Convention, "arid areas", may have to be replaced by the "marginal lands" concept. The term "marginal lands" would encompass ecosystems which are most susceptible to disturbances such as intense land use. In Iceland, deforestation and grazing are major causes for desertification, ecosystem degradation, and ultimately the creation of extensive barren wastelands. The parallels of desertification causes and processes between drier regions and the more humid Iceland are striking. The difference is the nature of the stress which makes these systems vulnerable to disturbance, i.e. dryness versus coldness. Too much rain in steep terrain, as exemplified by the effects of hurricane Mitch in Honduras in 1998, and volcanic eruptions, are other "natural" stress factors that interact with intensive land use to cause extreme land degradation.

Another point to consider is that ecosystem function is as strongly dependent on the fate of water on the ground as on overall rainfall. If the soil loses its ability to store water and supply it to plants, water shortages will be intensified in any climatic zone and contribute to degradation (Thurow, this volume; Imeson and Cammeraat, this volume). In Iceland, this scenario is taken to an extreme: erosion processes remove rich Andosols with high water holding capacity, replacing them with sandy soil surface with limited ability to store water and nutrients. As a result, water shortage becomes a severe limitation in spite of a humid climate.

Many authors have pointed out the importance of resilience in relation to desertification (e.g., Archer and Stokes, this volume; Tongway and Hindley, this volume). Warren and Agnew (1988) stated that "*degradation occurs when resilience is damaged* " and Rubio (1995) stressed that climatically extreme regions are more sensitive due to lesser resilience. When ecosystem resilience of marginal lands is reduced by the use of the land by man, the system's capacity to absorb stress or to respond when released from stress is diminished. This can be brought on by natural fluctuations in environmental stresses, such as drought and cold spells, and accelerated by human land use pressure. A new steady state may be reached with reduced productivity or a major collapse brought on resulting in near barren land (see Archer and Stokes, this volume).

4.3 Alternatives

There are at least two plausible alternatives to the current scope of the UN-CCD as determined by CCD definition of desertification. One is to change or improve the definition, the other is to broaden the context of the Convention.

The first alternative involves the adoption of a simple and open definition such as "degradation causing long-term reduction in the productivity of the land". Similar definitions have been suggested by Biot (1993) and by the EU DeMon project (Hill, 1996). Desertification would, in this context, be a broad scientific, social and political concept. More detailed definitions could be made for specific purposes, including local or regional assessments of the problem.

The other alternative is to change the scope of the Convention from desertification to severe land degradation with lasting effects on productivity. Marginal lands, such as arid, cold, steep terrain, and other fragile ecosystems would be most vulnerable to such degradation. This is no easy task, as international conventions are complex and involve long and tedious negotiations before agreements are reached. This alternative may be difficult, but merits discussion.

Severe land degradation is a global environmental problem. It is not constrained by political or climatic boundaries. Land degradation in humid, mountainous areas can cause massive desertification in drier lowlands because of poorer quality of irrigation water. Severe land degradation needs to be dealt with at a global level regardless of climatic boundaries. If neither of the alternatives outlined above are used, it is quite possible that it will be deemed necessary to develop a new international convention dealing with land degradation in general. That could greatly limit the success of the current CCD. This further underlines the need for a critical investigation of CCD conceptual problems and possible alternatives.

5. CONCLUSIONS

"Desertification" is a vague concept because the term 'desert' can have many different meanings. The recent UN definition (CCD, 1994), land degradation within arid areas, has a broad acceptance, but is subjected to severe limitations. As such, the definition excludes many areas of the world subjected to severe land degradation, for example India and SE-Asia, Central and much of South America. It puts an unnecessary narrow perspective on the problem which has damaging effect on international, social, political, and scientific communication in this field. Evolution of the CCD convention from the currently regionally limited concept towards a more comprehensive, robust convention, embracing all severe land degradation, is desirable. Such evolution of the CCD convention would enhance communication, research, and encourage counter measures at the global level.

REFERENCES

Allaby, M 1985. The Oxford Dictionary of Natural History. Oxford, New York.

Anderson, F.W. and Falk, P. 1935. Observations on the ecology of the Central Desert of Iceland. J. of Ecol. 23, 406-421.

Arnalds, A. 1987. Ecosystem disturbance in Iceland. Arctic and Alpine Res. 19, 508–513.

Arnalds, O. 1990. Characterization and erosion of Andisols in Iceland. Ph.D. dissertation, Texas A&M University, College Station, Texas.

Arnalds, O. 1998. Desertification in Iceland. Desertification Contr. Bull. 32, 22–24.

Arnalds, O. 1999. The Icelandic "rofabard" soil erosion features. Earth Surface Processes and Landforms. (In press).

Arnalds, O., Aradottir, A.L. and Thorsteinsson, I. 1987. The nature and restoration of denuded areas in Iceland. Arctic and Alpine Res. 19, 518–525.

Arnalds, O., Hallmark, C.T. and Wilding, L.P. 1995. Andisols from four different regions in Iceland. Soil Sci. Soc. Am. J. 59, 161–169.

Arnalds, O, Thorarinssdóttir, E.F., Metusalemsson, S., Jonsson, A., Gretarsson, E. and Arnason, A. 1997. Soil Erosion in Iceland. The Soil Conservation Service and Agricultural Research Institute, Reykjavik, Iceland.

Auberville 1949. Climats, forets et désertification de l'Afrique tropicale. Soc d'éditions géographiques et coloniales. Paris.

Behnke, R.H. and Scoones, I. 1993. Rethinking range ecology: Implications for rangeland management in Africa. In: Range Ecology at Disequilibrium. New Models of Natural Variability and Pastoral Adaptation in African Savannas. Eds. R.H. Behnke, I. Scoones and C. Kerven. pp. 1–30. Overseas Development Institute, London.

Bergthorsson, P. 1969. An estimate of drift ice and temperature in Iceland in 1000 years. Jokull 19, 94–101.

Binns, T. 1990. Is desertification a myth? Geogr. 75, 106–113.

Biot, Y. 1993. How long can high stocking densities be sustained. In: Range Ecology at Disequilibrium. New Models of Natural Variability and Pastoral Adaptation in African Sa-

vannas. Eds. R.H. Behnke, I. Scoones and C. Kerven. pp. 153–172. Overseas Development Institute, London.
Bolle, H.L. 1995. Climate and desertification. In: Desertification in a European Context: Physical and Socio-economic Aspects. Eds. R. Fantechi, D. Peter, P. Balabanis and J.L. Rubio. pp. 15–26. European Commission, DG-XII, Brussels.
CCD 1994. United Nations Convention to Combat Desertification in thouse Countries. Experiencing Serious Drought and/or Desertification, Particularly in Africa. UNEP, Nairobi, Kenya. Printed 1995 in Switzerland.
Cooke, R., Warren, A. and Goudie, A. 1993. Desert Geomorphology. UCL Press, London.
Dale, V.H. 1997. The relationship between land-use change and climate change. Ecol. Applic. 7, 753–769.
Einarsson, Th. 1963. Pollen analytical studies on the vegetation and climate history of Iceland in Late and Post-Glacial times. In: North Atlantic Biota and Their History. Eds. A. Löve and D. Löve. pp. 355–365. Pergamon Press, Oxford.
Forse, B. 1989. The myth of the marching desert. New Scientist February 4, 1989, 31–32.
Gisladottir, G. 1998. Environmental Characterisation and Change in South-western Iceland. Department of Physical Geography, Stockholm University Dissertation Series 10, Stockholm, Sweden.
Glantz, M.H. and Orlovsky, N.S. 1983. Desertification: A review of the concept. Desertification Contr. Bull. 9, 15–22.
Gudbergsson, G. 1975. Soil formation in Skagafjordur, northern Iceland. J. Agr. Res. in Iceland 7, 20–45. (In Icelandic, extended English summary).
Hallsdottir, M. 1995. On the pre–settlement history of Icelandic vegetation. Icelandic Agricultural Sciences 9, 17–29.
Helldén, U. 1991. Desertification – time for an assessment? Ambio 20, 372–383.
Hill, J. 1996. DeMon. Intergrated Approaches to Desertification Mapping and Monitoring in the Mediterranean Basin. European Commission, DG-XII, Brussels.
Kristinsson, H. 1995. Post-settlement history of Icelandic vegetation. Icelandic Agricultural Sciences 9, 31–35.
LMI 1993. Digital Vegetation Map of Iceland. The Icelandic Geodetic Survey, Reykjavik, Iceland.
Magnusson, S.H. 1994. Plant Colonization of Eroded Areas in Iceand. Ph.D. thesis. Lund University, Sweden.
Mainguet, M. 1994. Desertification. Natural Background and Human Mismanagement. 2nd edn. Springer-Verlag, Berlin.
Newman, J.R. 1967. The Harper Encylopedia of Science. Harper and Row, Washington DC.
Odingo, R.S. 1990. The definition of desertification and its prgrammatic consequences for the international community. In: Desertification revisited. Ed. R.S. Odingo. pp. 7–44. UNEP/DC-PAC, Nairobi, Kenya.
Pearce, F. 1992. Mirage of the shifting sands. New Scientist December 12, 1992, 38–42.
Perez-Trejo, F. 1992. Desertification and Land Degradation in the European Mediterranean. EC, DG-XII, Brussels.
Rieger, S. 1983. The Genesis and Classification of Cold Soils. Academic Press, New York.
Rubio, J.L. 1995. Opening conference – Desertification: evolution of a concept. In: Desertification in a European Context: Physical and Socio-economic Aspects. Eds. R. Fantechi, D. Peter, P. Balabanis and J.L. Rubio. pp. 5–13. European Commission, DG-XII, Brussels.
Sigbjarnarson, G. 1969. The loessial soil formation and the soil erosion on Haukadalsheiði. Natturufraedingurinn 39, 49–128. (In Icelandic, extended English summary).
Sigurdsson, S. 1977. Birch in Iceland. In: Skogarmal. Ed. H. Gudmundsson. pp. 146–172. Edda Print, Reykjavik, Iceland. (In Icelandic).

Thomas, D.S.G. and Middleton, N.J. 1994. Desertification: Exploiding the Myth. John Wiley & Sons, New York.

Thorarinsson, S. 1961. Wind erosion in Iceland. A tephrocronological study. In: Icelandic Forestry Society Yearbook 1961, pp. 17–54. (In Icelandic, extended English summary).

Thorarinsson, S. 1981. The application of tephrocronology in Iceland. In: Tephra Studies. Eds. S. Self and R.S.J. Sparks. pp. 109–134. D. Reidel, London.

Thornes, J.B. 1995. Mediterranean desertification and the vegetation cover. In: Desertification in a European Context: Physical and Socio-economic Aspects. Eds. R. Fantechi, D. Peter, P. Balabanis and J.L. Rubio. pp. 169–194. European Commission, DG-XII, Brussels.

UNEP-DC/PAC 1990. In: Desertification revisited. Ed. R.S. Odingo. pp. 3–5. UNEP/DC-PAC, Nairobi, Kenya.

UN 1977. Desertification, its Causes and Consequences. Pergamon Press, Oxford.

Warren, A. and Agnew, C. 1988. An Assessment of Desertification and Land Degradation in Arid and Semi-arid Areas. IIED International Institute for Environment and Development Drylands paper no. 2. London.

Whittaker, R.H. 1975. Communities and Ecosystems. MacMillan Publishing Co., New York.

Whitten, D.G.A. and Brooks, J.R.V. 1974. A Dictionary of Geology. Penguin, London.

Stress, disturbance and change in rangeland ecosystems

Steve Archer and Chris Stokes
Department of Rangeland Ecology & Management, Texas A&M University, College Station, TX 77843-2126, USA
Tel: 409 845 7332; Fax: 409 845 6430; E-mail: <sarcher@vms1.tamu.edu> & <cstokes@tamu.edu>

ABSTRACT Ecological systems and the organisms which comprise them have evolved with and are a product of various stresses, perturbations, and disturbance regimes. However, in human-influenced systems, new disturbances and stresses may be introduced and the frequency, intensity and spatial extent of natural disturbances altered. Natural and anthropogenic disturbances invariably co-occur, so it becomes difficult to ascertain which may be the proximate cause of ecosystem change. It is likely that their effects are compounded by synergistic interactions. In some cases, anthropogenic disturbance may alter the susceptibility of organisms, populations and communities to natural disturbances. In other cases, anthropogenic activities may initiate positive feedbacks that produce rapid, unexpected changes in ecosystem structure and function. These changes may be stabilized by new ecosystem processes, making them irreversible over time frames relevant to management.

An understanding of stress and disturbance will help resource managers to (1) mitigate anthropogenic disturbances which might threaten sustainability and lead to undesirable and potentially irrevocable changes in ecosystem processes and (2) increase chances for success in rehabilitating or restoring degraded ecosystems. Here, we review the role of stress and disturbance in regulating the structure and function of rangeland ecosystems, and present conceptual models of ecosystem change which result from alterations of disturbance regimes. We then discuss rates and dynamics of change and present examples which illustrate how anthropogenic alteration of natural disturbance regimes can alter ecosystem stability and produce multiple stable states. We conclude with a brief discussion of land degradation from ecological vs. socio-economic perspectives.

Key words: degradation, feedbacks, resilience, stability, thresholds, transitions.

1. INTRODUCTION

Degradation and desertification processes are typically associated with arid and semi-arid regions. However, this represents a rather narrow view

(O. Arnalds, this volume) and ignores important changes in land cover which occur in more humid environments (Fig. 1).

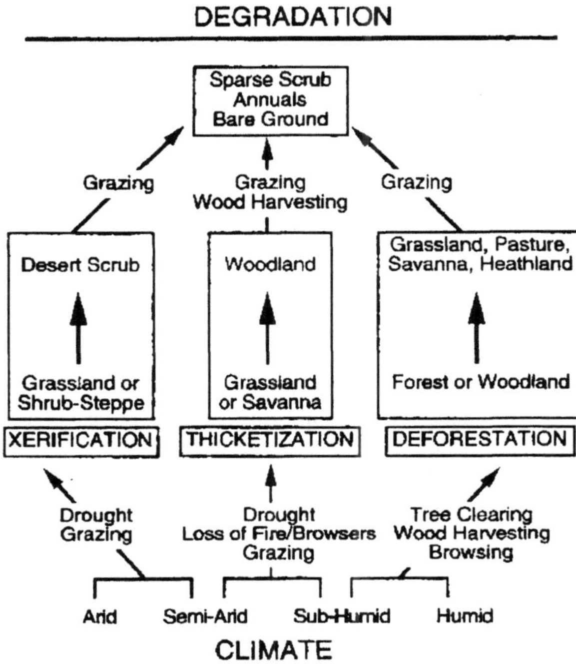

Figure 1. Degradation may take many forms and proceed via different pathways, depending upon climate and anthropogenic modifications of natural disturbance regimes. Although 'desertification' is commonly associated with arid and semi-arid regions, its functional equivalent can occur in high rainfall areas as well. Narrow definitions of 'desertification' therefore exclude other cases of severe degradation and hinder progress in research, management and policy (O. Arnalds, this volume). For 'xerification' see West (1986) and Havstad (this volume); for woodland deforestation, see Gadgill and Meher-Homji (1985), Reid et al. (1990) and O. Arnalds (this volume); thicketization is discussed later in this chapter.

It is widely acknowledged that cropping and intensive livestock grazing may result in a loss of vegetation cover which contributes to erosion and desertification (Thurow, this volume). However, in other instances, intensification of livestock grazing and reductions in fire frequency have altered grass-woody plant interactions in favor of unpalatable trees and shrubs, resulting in the transformation of grasslands and savannas into shrublands (Schlesinger et al., 1990) and woodlands (Archer, 1994). While such changes may be regarded as 'degradation' from a socioeconomic perspective (e.g., reduced carrying capacity for livestock) do they represent 'ecological degradation' (i.e., a loss of biodiversity, primary production, or nutrient capital)? Extensive tree clearing in more mesic environments, has created savannas in tropical/subtropical regions (Gadgill and Meher-Homji, 1985), shrublands in Mediterranean regions (Yassoglou, 1999) and heathlands in high latitude regions (A. Arnalds, 1987; O. Arnalds, this volume). These derived systems may be productive in the short term, but subsequently degraded over the longer term. Because degradation and desertification-like processes span such a broad gamut of climatic settings and ecosystem physiognomies, it is difficult to develop meaningful generalizations or robust conceptual models. Ludwig and Tongway (this volume) present a conceptual framework for viewing spatial processes associated with desertification and degradation of ecosystems and landscapes. Our chapter emphasizes rates and dynamics of change, the interactive effects of natural and anthropogenic stress and disturbance, susceptibility to change and the prognosis for ecosystem recovery.

2. DISTURBANCE AND STRESS IN RANGELAND ECOSYSTEMS

The terms 'disturbance', 'perturbation' and 'stress' have been used inconsistently and ambiguously. Defining these terms is difficult, because ecological systems are probably always in a non-equilibrium state at some spatial/temporal scale as they are continuously adjusting to fluctuating abiotic (e.g., temperature, rainfall) and biotic (competition, predation) conditions. Disturbances are only abnormal in the sense that they are not continuous and that they cause an ecosystem characteristic such as species diversity, nutrient output, or biomass to exceed or drop below the homoeostatic range of variation regarded as normal (Godran and Forman, 1983). However, our recognition of what constitutes 'normal' variation is typically qualitative, value-laden and relative to reference states which may be arbitrary or ill-defined. As a result, quantifying deviation from 'normal' or the degree to which a system has returned to 'normal' following the alleviation of stress or disturbance is difficult.

Disturbances may be regarded as biotic and abiotic forces, agents or processes which cause perturbations and induce stresses (Rykiel, 1985). Disturbances are categorized and quantified with respect to type (e.g., grazing, fire, flood and wind), frequency (common vs. episodic), extent (patch/ local-scale vs. landscape/ regional-scale) and intensity (e.g., fires may burn hot or relatively cool). Disturbances may cause (a) the destruction of biomass; (b) the selective elimination, reduction, addition or expansion of populations; (c) disruption of matter, energy or information exchange; and/or (d) the suppression of another disturbance. Grazing, for example, is a disturbance that removes grass biomass aboveground and curtails root production below-ground. If maintained at high intensity for sufficiently long periods, grazing can lead to a loss of plant cover, shifts in species composition, erosion or volatilization losses of soil nutrients, and the elimination of natural disturbances such as fire. A disturbance regime is the sum of types, frequencies and intensities of disturbance through time in the landscape. Frequent, small-scale disturbances, such as those resulting from patch grazing, wallowing, and urine and dung deposition by herbivores contribute to the development of fine-grained mosaics across landscapes which may experience infrequent fire and rare, but recurring, drought which affect the entire landscape and watershed. Interpreting ecosystem structure and function is thus contingent upon understanding the interactive role of concurrent, multiple-scale disturbances (Collins and Barber, 1985; Loucks et al., 1985; Chapin et al., 1987; Collins, 1987).

Disturbances are generally perceived as negative, because they are accompanied by elements of damage. However, a disturbance at one level of organization does not necessarily induce change if it is 'absorbed' at higher levels of ecological organization. Disturbance can therefore be viewed as part of the internal workings of an ecosystem rather than as exogenous events (see Ludwig and Tongway, this volume). Indeed, some level of disturbance may be necessary for maintenance of ecosystem properties. Savanna landscapes, for example, are often stable despite consisting of numerous patches in various states of transition between grass and tree dominance caused by grazing, browsing and fire (Dublin et al., 1990; Scholes and Archer, 1997).

Perturbations can be viewed as the response of an ecological component or system to disturbance, as indicated by deviations relative to a specified reference condition (Rykiel, 1985). Perturbations are transient if the system returns to its original steady state or permanent if they lead to a deviation which becomes fixed and produces a steady state different from the original.

Stresses are specific factors that adversely affect the physiology or function of organisms and ecosystems. Factors which alter resource availability or which push resource modulators such as temperature to extremes induce stress. Stresses which result from disturbance or weather may be chronic (relatively continuous, but at low levels) or acute (having a rapid onset and followed by a short, but severe effect) and may either impair function or cause mortality. Stresses do not typically occur in isolation (Chapin et al., 1987).

3. DISTURBANCE, CLIMATE AND STRESS INTERACTIONS

Disturbances are superimposed on a background of topoedaphic heterogeneity and climatic variability. As a result, plant species whose adaptations to the prevailing climate and soils would make them the competitive dominants of the community under one disturbance regime, may assume subordinate roles or even face local extinction when disturbance regimes change. Disturbances such as fire, flooding, and pathogen outbreaks may be directly or indirectly related to climate (e.g., Swetnam and Betancourt, 1990).

In many instances it is difficult to determine the extent to which climate change and disturbance are responsible for driving ecosystem changes (Herbel et al., 1972; Chew, 1982; McNaughton, 1983; Branson, 1985; Foran, 1986; Verstraete, 1986). The difficulty in making this distinction is, in part, due to the fact that shifts in vegetation may lag well behind the climatic changes that drive them. For example, it has been hypothesized that the desert grassland vegetation of the Chihuahuan Desert recorded by settlers of North America in the late 1800s, established under and adapted to 300 years of cooler, moister climates of the 'Little Ice Age' and is only marginally supported under the present climate (Neilson, 1986). If this hypothesis is correct, climate-driven 'desertification' of these grasslands and their replacement by desert scrub may have been in progress at the time of Anglo-European settlement, and augmented, but not caused, by anthropogenic alteration of grazing and fire regimes.

Disturbance and climate can interact to offset or reinforce ecosystem change. Disturbance effects may be minimal when climatic conditions are favorable or magnified when climatic conditions are extreme. Anthropogenic activities may reinforce or accelerate changes triggered by natural events. For example, grassland retrogression associated with livestock grazing may be mitigated in years of normal or above-normal rainfall and magnified during years of below-normal precipitation (Herbel et al., 1972; Clarkson and Lee, 1988; O'Connor, 1993; Orr et al., 1993; O'Connor, 1994). The encroachment of unpalatable trees and shrubs into grazed subtropical grasslands and savannas illustrates the potential for interaction between climate and grazing disturbance (Fig. 2). In this case, woody plant cover decreased slightly between 1941 and 1960, apparently the result of a major drought during the 1950s. In the subsequent pluvial period, woody plant cover increased 2- to 4-fold. Changes in ecosystem structure have therefore been punctuated and abrupt rather than gradual and continuous. Drought may have predisposed the system to rapid rates of woody plant invasion in the post-drought period. However, these sites had also been heavily grazed by cattle since the mid-1800s. Cattle effectively disperse seeds of the dominant woody plant and it is likely that their utilization of grasses intensified drought stresses on these plants. Continued post-drought grazing would also have limited the mass and continuity of fine fuels needed to carry fire. Livestock grazing could thus contribute significantly to woody plant expansion. Would either drought or livestock grazing alone have been sufficient to produce this pattern and magnitude of change? It is likely that these changes were the result of the combined effects of these two disturbances acting together. Thus, certain changes in environmental conditions may be necessary, but they are not sufficient by themselves to elicit a change in ecological systems.

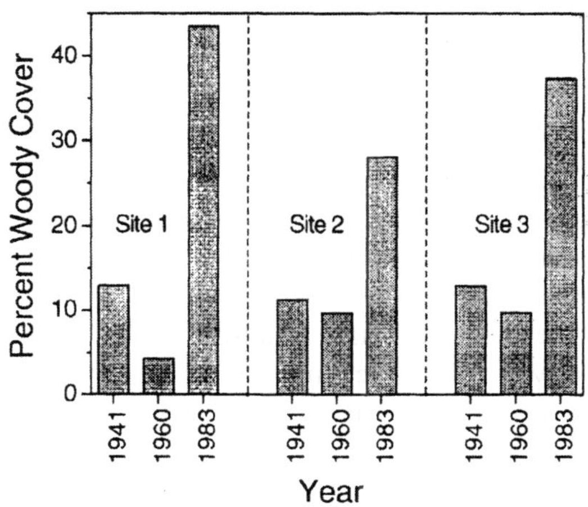

Figure 2. Changes in landscape cover of woody vegetation in a subtropical savanna parkland between 1941 and 1983 (from Archer et al., 1988). This site has a history of heavy, continuous grazing by cattle and experienced a major drought during the 1950s. It is likely that drought and grazing combined to affect the dynamics of bush encroachment in a way that neither factor operating independently would have. See also Case III in Fig. 6.

Disturbances serve to overcome vegetation inertia by lessening the dominance of established plants and creating opportunities for new plants which may be better adapted to the present climatic conditions. Ecosystem simulations indicate that climate change alone may have a minimal short-term influence on composition. For example, in simulations by Overpeck et al. (1990), the response of forest vegetation to climate change took >200–250 years to achieve equilibrium in the absence of disturbance. By contrast, vegetation subjected to disturbance closely tracked the initiation and timing of climate change and produced a vegetation with novel and unique characteristics.

4. MULTIPLE STEADY STATES, TRANSITION THRESHOLDS AND POSITIVE FEEDBACKS

Once ecosystem change has occurred, relaxation of stress and disturbance or an improvement of environmental conditions will not necessarily enable a system to return to its previous state (c.f., Rapport and Whitford, 1999). Alteration of soil structure (Thurow, this volume), distribution and abundance of water and nutrients (Ludwig and Tongway, this volume; Havstad et al., this volume; Tongway and Hindley, this volume), and plant composition (Hobbie, 1992) may occur and re-direct ecological processes and stabilize the site in a new steady-state arrangement. This suggests the existence of critical transition thresholds. Stochastic, climatic events may hasten disturbance-mediated transitions (e.g., O'Connor, 1993). Transitions may be further accelerated if "keystone species" establish and initiate positive feedbacks to redirect succession or change disturbance regimes (Archer et al., 1988; D'Antonio and Vitousek, 1992).

How far can ecological systems be pushed before crossing the line of ecological function that separates one steady-state from another? Are contrasting ecosystem states endpoints of a continuum of steady, continuous change? Or, is a given ecological state relatively stable over a range of disturbance regimes and environmental conditions, but then prone to radical change when these ranges are exceeded? The former implies gradual, incremental change; the latter a highly discontinuous behavior. As suggested in the following examples, both are likely to occur and their relative importance is often a matter of perspective (see also Fig. 4 in Tongway and Hindley, this volume). Detection and definition of thresholds will depend on values attached to signal and noise in ecological processes, the spatial and temporal frames of reference, the purpose behind studying ecosystem change, and the indicators used to quantify ecosystem condition and variation.

4.1 High latitude example

Desertification of Icelandic rangelands has been extensive and spectacular (A. Arnalds, 1987; O. Arnalds, this volume). The degradation process (Figs. 3 and 4) appears to begin when unregulated livestock grazing shifts vegetation from a continuous cover of palatable deciduous shrubs, grasses and forbs (State I) to less productive heathland dominated by unpalatable evergreen dwarf shrubs and forbs (State II) (Aradottir et al., 1992). Reduction of the plant biomass thermal barrier makes soils more prone to frequent small-scale disturbances associated with frost boils and frost heaving. This hastens the formation of bare patches, which expose the friable, thick (50–200 cm) mantle of volcanic soil (State III). The exposed soil mantle is then removed by wind and water erosion (O. Arnalds, 1998). As these eroded patches enlarge and coalesce, the length of exposed, eroded perimeter increases dramatically, creating 'erosion fronts' whose vertical faces (rofabards) are fully exposed to wind (O. Arnalds, 1990, 1999). These elongated, wind-driven fronts can now advance rapidly across the landscape, leaving an infertile soil surface in their wake. Management of the remaining vegetated zones does little to prevent the advance of the erosion fronts. Self-reinforcing changes now occur very rapidly (States III and IV), culminating in a landscape characterized by either glacial till or sandy surfaces with widely scattered vegetated remnants (States V and VI).

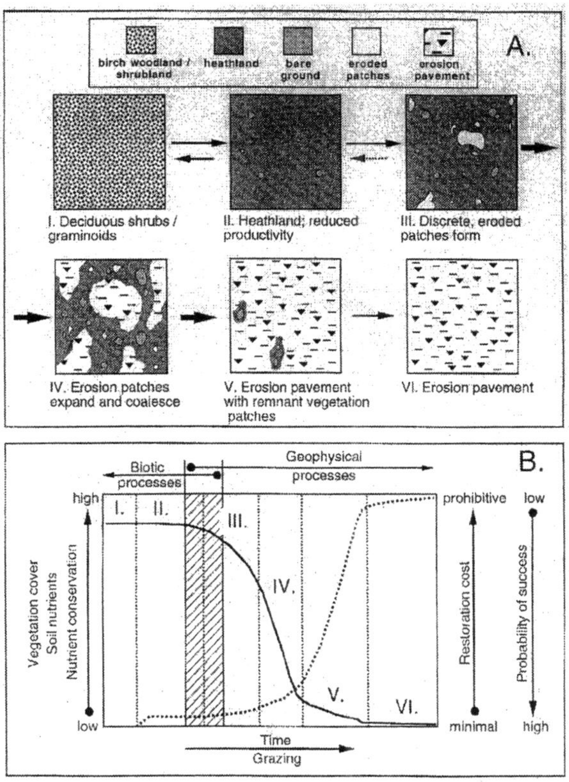

Figure 3. Degradation of birch woodlands in Iceland (adapted from Aradottir et al., 1992). (A) Changes in land cover status brought about by grazing and erosion. (B) Conceptual model of the dynamics of change in land cover status (solid line) and associated restoration costs (dashed line). (For further elaboration see Figs. 3 and 4 in Tongway and Hindley, this volume).

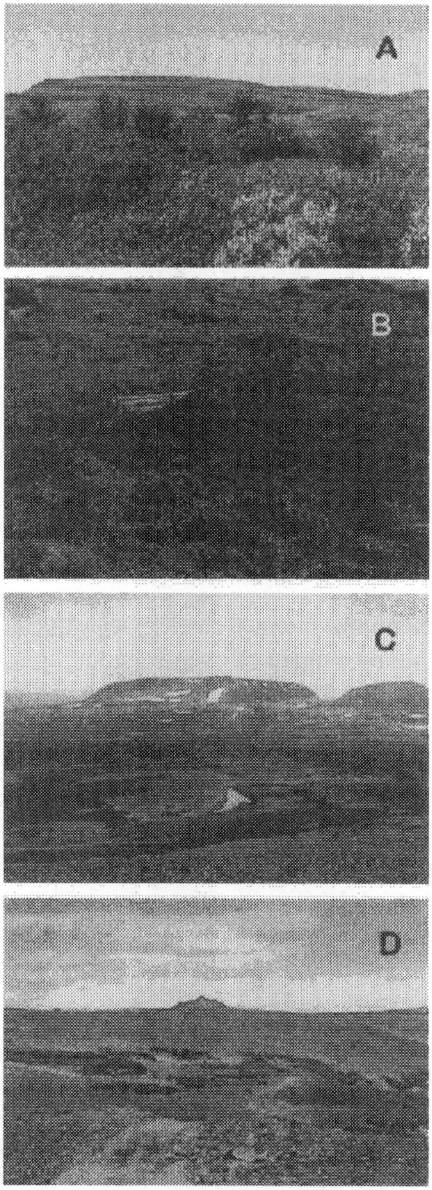

Figure 4. Icelandic landscape states represented in Fig. 3. (A) Birch (*Betula pubescens*) shrubland typical of areas protected from sheep grazing; (B) Heathland vegetation associated with grazed areas. Note bare ground patches. These sites are now increasingly destabilized by frost heaving and are subject to wind/water erosion (C) Erosion escarpments (rofabards) that move rapidly across the landscape regardless of vegetative cover. (D) Glacial till and remnant soil. Aeolian deposition of soils from these sites create active dunes that bury vegetation and soils on other sites.

Changes appear to be rather gradual and continuous from State I to State II. Soil structure, fertility, plant cover and biotic processes which modulate climate and nutrient fluxes are still in place. Thus, relaxation of grazing and progressive intervention may permit a return to State I. The system at State I may also be resistant to environmental stresses, such as unusually cold growing seasons or tephra deposition associated with volcanic activity. However, in State II, these same stresses may push the system past a threshold, propelling it to states of advanced degradation. For landscapes in State III, relaxation of grazing (i.e., removal of disturbance) may be of little consequence, because geophysical processes (frost heaving and erosion) alone are driving the system. State VI is highly stable and characterized by geophysical forces and ecological processes very different from those in preceding states. Once in these degraded states, management and restoration options are expensive to implement and have a high probability of failure.

4.2 Semi-arid savanna example

In the Icelandic example, interactions between livestock grazing and environmental events initiate a chain of events leading to a total loss of vegetative cover and massive erosion. In cool and warm temperate arid zones, replacement of grasses by unpalatable shrubs, also potentially initiated by grazing, may cause a redistribution of nutrients and an overall loss of primary and secondary productivity (Kieft et al., 1998; Havstad et al., this volume). These examples contrast starkly with land degradation in sub-humid tropical and subtropical regions, as described in the ensuing paragraphs. Even so, conceptual models of vegetation change based on resilience and thresholds appear applicable.

Numerous quantitative assessments corroborate historical observations which indicate that many regions of North and South America, Africa and Australia were characterized by grasslands and savannas at the time of Anglo-European settlement. Many such areas are now woodlands or shrublands, dominated by unpalatable trees and shrubs (reviewed by Archer, 1994, 1996). These changes in vegetation cover are widely regarded as degradation from a socioeconomic perspective because they have reduced the potential of the land for livestock production.

In semi-arid and sub-humid grasslands and savannas, tall and mid-height grasses are typically the competitive dominants under light grazing; woody plant density is low and grass competition and periodic fire suppress their recruitment. Short-statured grasses are more tolerant to defoliation than the taller grasses or are less preferred. As a result, they increase in abundance with intensification of grazing (Archer and Smeins, 1991). Even so, ground cover remains high, soils and microclimate are relatively unaffected, and the amount and spatial distribution of fine fuels permits natural or prescribed fire in many years. With relaxation of grazing, taller grasses regain dominance of the site via regeneration from seed or vegetative propagules (Fig. 5).

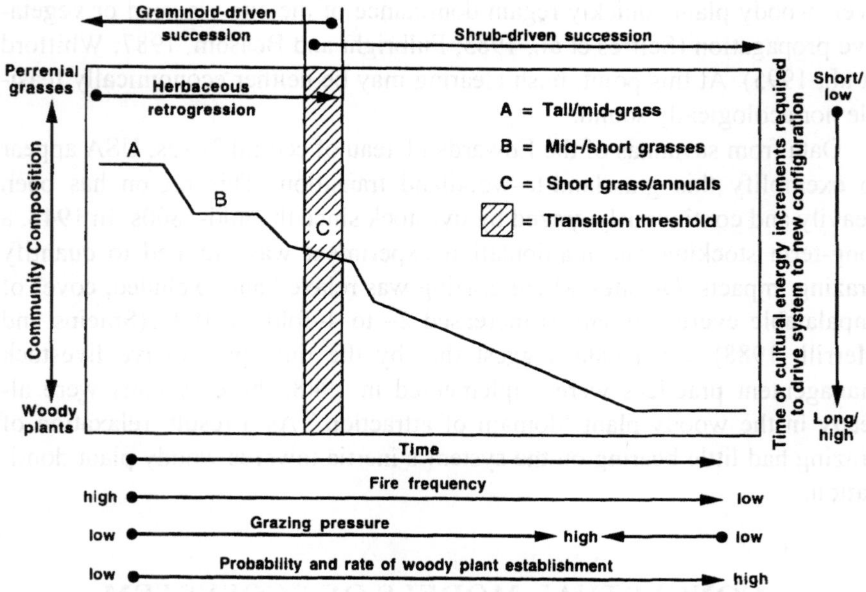

Figure 5. Conceptual model of 'thicketization' (Fig. 1) and transition thresholds in a semi-arid, subtropical savanna (from Archer, 1989). The goal of research should be to characterize plant population or edaphic properties that may forecast impending transitions. The goal of management should be to adjust land use practices to avert crossing thresholds leading to undesirable and potentially irreversible changes in vegetation or soils. (See also Tongway and Hindley, this volume; Ludwig and Tongway, this volume).

With continued intensification of grazing, perennial short-grasses are replaced with annuals and weakly perennial grasses of low litter and nutrient quality. Bare ground increases and promotes warmer, drier microclimates and wind or water erosion (Thurow, this volume). Fire may be virtually eliminated, as there is seldom sufficient fine fuel for ignition and spread (Baisan and Swetnam, 1990; Savage and Swetnam, 1990). Unpalatable N_2-fixing or evergreen shrubs adapted to the harsher microclimate and low nutrient environments may now establish. As woody plant density increases, herbaceous production declines further, thus increasing the grazing pressure on remaining herbaceous patches. By this time, a physiognomic or domain threshold has been crossed, whereby soils, seedbanks and vegetative regenerative potentials have been modified such that succession will now progress toward shrubland or woodland even if grazing pressure is relaxed. The system has therefore moved from a grassland domain to a shrubland or woodland domain. Drought, natural disturbance, or anthropogenic manipulation (chemical/mechanical brush management practices) may open up woody plant canopies and enable grass production to increase temporarily. However, woody plants quickly regain dominance of the site via seed or vegetative propagation (Scifres et al., 1983; Fulbright and Beasom, 1987; Whitford et al., 1995). At this point, bush clearing may be neither economically feasible nor ecologically sound.

Data from savannas of the Edwards Plateau of central Texas, USA appear to exemplify this grassland-to-woodland transition. This region has been heavily and continuously grazed by livestock since the mid-1800s. In 1948, a long-term stocking rate manipulation experiment was initiated to quantify grazing impacts. On sites where grazing was relaxed and excluded, cover of unpalatable evergreen shrubs increased 2– to 4–fold by 1983 (Smeins and Merrill, 1988). Such data suggest that by the time progressive livestock management practices were implemented in 1948, these systems were already in the woody plant 'domain of attraction.' As a result, relaxation of grazing had little bearing on the system's inertia towards woody plant domination.

5. CONCEPTUAL MODELS OF ECOSYSTEM CHANGE

Traditional 'retrogression sequence' models have dealt with single transitions involving responses to particular disturbance regimes (e.g., Dyksterhuis, 1949; Bosch and Booysen, 1992). These models have been continually refined to deal with inadequacies in the successional theory on which they were originally based (Connell and Slatyer, 1977; Archer, 1989; Luken, 1990; Friedel, 1991; Laycock, 1991; Joyce, 1993). Another approach has been to consider the responses of ecosystems to different sets of stress and disturbance regimes and the alternate ecosystem states that result from these transitions (George et al., 1992; Jones, 1992; Milton and Hoffman, 1994). These 'state and transition' (S&T) models have also been useful for understanding the dynamics of vegetation mosaics in highly stochastic environments (Westoby et al., 1989b; Hobbs, 1994; Hoffman, this volume).

State and transition models have gained wide support, but also have limitations (Watson et al., 1996). The definitions of states are typically descriptive 'black boxes' and our understanding of the factors conferring stability within a state or of factors triggering transitions between states is limited. Similarly, the rates and dynamics at which a system moves from one state to another are seldom known. Here, we elaborate a conceptual framework for exploring the dynamics of transitions between states (i.e., the 'arrows' in S&T diagrams).

Long-term, sustained stresses such as grazing can act as directional forcings for gradual, incremental state transitions. However, episodic (rare, but recurring) perturbations in stochastic environments can produce rapid transitions with unpredictable outcomes. Although dogma from dryland systems has emphasized 'event-driven' change, gradual, continuous changes may also be of comparable importance (Watson et al., 1997a,b). Both episodic and continuous changes are therefore of potential consequence. When stresses exceed the buffering capacity of the system, there will be a net imbalance between degenerative and regenerative processes (disequilibrium) leading to a deviation from the initial ecosystem state. These stresses may be natural or anthropogenic. The synergistic interactions of excess stresses and intrinsic disturbance regimes can be conceptualized as a 'pump and valve' system. Periodic environmental stress and disturbance act as a pump for ecosystem change. Unbalanced stresses then act as 'valves' that may favor degenerative changes over processes of regeneration. The greater the stress, the greater the potential for rapid, discontinuous change triggered by stochastic, abiotic forces (e.g., Lockwood and Lockwood, 1993).

There are several mechanisms by which chronic and episodic stress and disturbance may interact to cause degradation (Fig. 6). These are ranked in order of increasing severity for degenerative ecosystem change (or, conversely, increasing desirability for processes of restoration and 'positive' ecosystem change) and increasing potential to cause a transition across a threshold between ecosystem states:

- **Steady-state fluctuations (no excess stress)**: In a 'healthy' ecosystem, which is not experiencing excess additional stresses, negative feedbacks maintain a steady state as environmental conditions (e.g., rainfall variability) oscillate around a long-term mean. Perturbations are fully buffered by biotic recovery processes. Fluctuations in ecosystem condition occur, but there is no net deviation from the steady state over the long-term. The ecosystem can be considered to be in dynamic equilibrium with its associated disturbance regime (at a spatiotemporal scale that encompasses fluctuations and patch dynamics).
- **Degradation I (suppressed regeneration)**: The effects of additional stresses (such as increased grazing pressure) may not be evident during unfavorable periods. Instead, they may act by limiting the potential for recovery during the regenerative phase of ecosystem fluctuations. Ecosystem structure may be relatively intact, but rates of functional processes may be slowed. For example, as soil properties (Thurow, this volume), seed production, vegetative regeneration or the seed bank are impacted,

system resilience is diminished. Full recovery, therefore, becomes increasingly unlikely, even with removal or alleviation of stresses. Some models of succession make allowance for the effects of disturbance in suppressing successional 'repair' processes (Schlatterer, 1989; Luken, 1990). This form of degradation is common where rangelands are being managed 'on the edge' of sustainability, with the goal of maximizing utilization and off-take. These levels of utilization may appear to be sustainable under 'average' climate conditions or during long runs of favorable years, but gradually deplete the reserve resources required to recover from periods of unfavorable conditions.

- **Degradation II (accentuated degeneration)**: In this scenario, the added stresses act to directly advance degenerative processes. The stresses exaggerate the impacts of adverse environmental conditions (e.g., drought) to the extent that the perturbations cannot be fully counteracted by processes of regeneration in the ensuing favorable period. This is the most basic concept of degradation, where the synergistic effects of natural and anthropogenic stresses overwhelm the ecosystem's buffering capacity.
- **Degradation III (degenerative 'recovery')**: In some cases, effects of the modified stress and disturbance regime may only be expressed during favorable periods. For example, disturbance may enable establishment of undesirable species whose recruitment is subsequently promoted during favorable periods. Or, preferential utilization of plants by livestock may reduce seed production or vegetative regeneration of desirable species, while enhancing that of unpalatable species (Milton, 1992; Stokes, 1994). Favorable periods of vegetation regeneration would then be biased to recruitment of the unpalatable species (see also Fig. 2). Undesirable states thus develop and are reinforced, becoming increasingly favored over time as a positive feedback develops. In the worst cases, positive feedbacks would be established in which the delayed response would continue long after the inducing stresses had been alleviated. This would indicate the crossing of a physiognomic or domain threshold with potentially irreversible changes to the ecosystem (Fig. 5). Establishment of species that change the disturbance regime can also initiate these undesirable positive feedbacks. Exotic grasses, for example, can increase fine fuel loads and hence the probability of fire. Favorable environmental periods enhance their establishment and biomass accumulation, thus increasing fire frequency and intensity to further promote their spread into new habitats (D'Antonio and Vitousek, 1992). Positive feedbacks dominated by abiotic processes (e.g., Icelandic example; Figs. 3 and 4) are potentially more devastating than positive feedbacks dominated by biotic processes (e.g., 'thicketization' of savannas; Figs. 2, 5 and 7). Biotic positive feedbacks may lead to a change in ecosystem composition and function,

whereas abiotic positive feedbacks are often accompanied by nudation and loss of basic ecosystem functions (e.g., nutrient cycling).

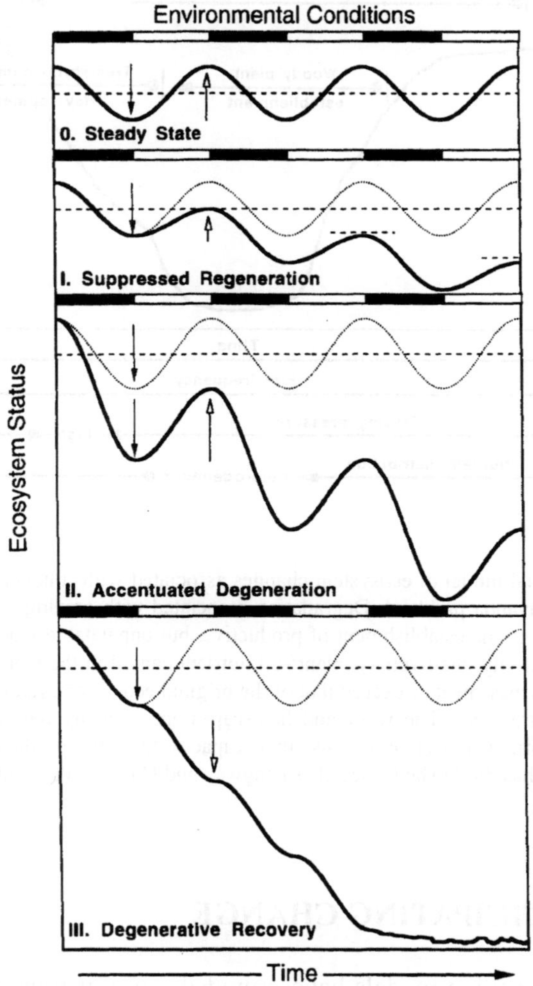

Figure 6. Conceptual model of ecosystem degradation in conjunction with climatic and environmental fluctuation. Dark arrows denote degenerative changes that occur during stressful environmental periods (solid segments of x-axis); clear arrows denote regenerative changes occurring during favorable environmental periods (clear segments on x-axis). Rates and dynamics of directional change (retrogression, degradation) are influenced by climatic variability and may be caused by dampening of recovery processes (Case I), accentuation of stress levels (Case II) and regeneration that is biased towards recruitment of undesirable species during favorable periods (Case III).

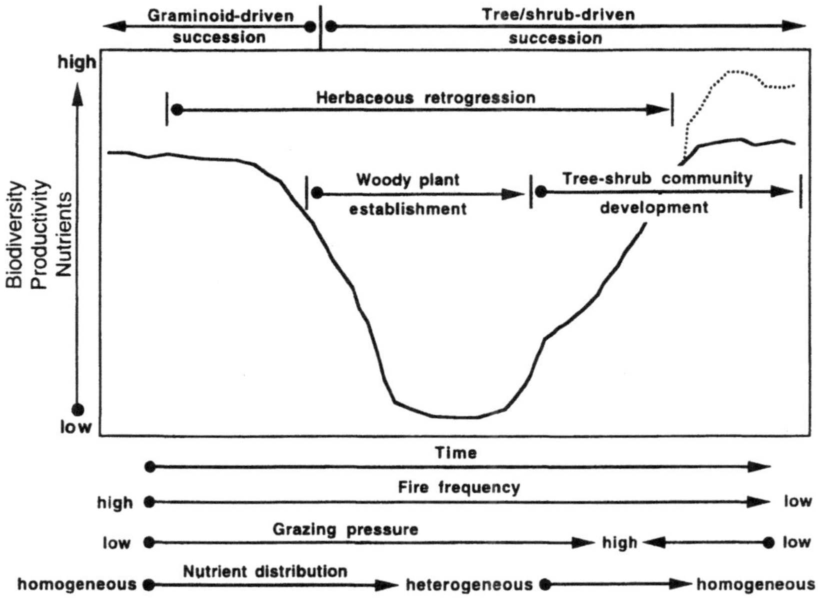

Figure 7. Conceptual model of ecosystem changes associated with 'thicketization' of a semi-arid, subtropical savanna parkland. Degradation associated with grazing-induced herbaceous retrogression leads to the establishment of productive, but unpalatable woody plants (Figs. 2 and 5). The productivity, biodiversity, plant/soil nutrient capital in the woodlands which subsequently develops may meet or exceed that of the original ecosystem. Ecological degradation may therefore be transient. The woodland landscapes are typically regarded as 'degraded' from a socioeconomic perspective, because their capacity for cattle production, the historical use of these lands, has diminished. (See also Tongway and Hindley, this volume).

6. ANTICIPATING CHANGE

State and transition models have provided a useful framework for highlighting critical events that create management hazards (unfavorable transitions) and opportunities for intervention (conditions for favorable transitions) (Westoby et al., 1989a). We now need a better quantitative and mechanistic understanding of why these transitions occur and how biotic and abiotic factors interact to promote or discourage them. From an applied perspective, we need to clarify how management actions before, during and after critical events might alter transition probabilities. A better understanding of transition mechanisms will also help identify and clarify key monitoring variables. Important changes that affect system response to perturbation events may go undetected unless variables with appropriate sensitivity are monitored (Stokes, 1994). Thus, there is an important need to develop sensitive indicators that can give advance warnings, so that ecosystem management can be adjusted to minimize the likelihood of undesirable transitions or to promote desirable transitions.

7. DEGRADATION: ECOLOGICAL OR SOCIOECONOMIC?

Degradation associated with 'desertification' in arid environments (e.g., Schlesinger et al., 1990; Rapport and Whitford, 1999; Havstad et al., 1999) or 'deforestation' in humid environments (Figs. 3 and 4) are a sharp contrast to that associated with the 'thicketization' of grasslands and savannas in more mesic environments. The former have negative consequences both ecologically and socioeconomically. Thicketization may have adverse socioeconomic implications, as it reduces the capacity of rangelands for subsistence or commercial livestock production. However, it does not necessarily represent a degraded system with respect to ecological characteristics such as biodiversity, productivity, or plant/soil nutrient capital.

The conceptual model in Fig. 7 indicates that a degradation phase may be followed by a reconstruction phase which begins when unpalatable woody plants establish, grow, modify microclimate and enrich soil nutrients. Dynamic simulations in subtropical savanna parklands in southern Texas support this conceptual model and indicate that degradation of grassland occurred subsequent to intensification of livestock grazing in the mid-1800s. This was manifested as a significant decline in primary productivity and plant/soil carbon and nitrogen pools. However, with selective grazing and the elimination of fire, nitrogen-fixing woody plants invaded, ameliorated microclimatic conditions, enriched soil nutrient pools and facilitated the establishment of a diverse assemblage of herbaceous dicots and shrubs (Archer, 1995). Present-day landscapes are a rich mosaic of productive woodlands and tree-shrub patches interspersed with remnant grass-dominated patches. It is estimated that current plant and soil C and N mass is substantially greater than that which occurred under 'pristine' conditions (Hibbard, 1995). In addition, these landscapes are highly resilient following severe disturbance (Flinn et al., 1992; see also Whitford et al., 1995) and are regarded as excellent habitat for numerous wildlife species, both game and non-game. So, in this case, the system which has developed following an initial degradation phase is now ecologically productive and diverse. It would seem that it is 'degraded' only with respect to its socioeconomic value for cattle grazing. However, it has other potential socioeconomic values that necessitate a change from traditional uses. These include alternate classes of livestock (e.g., goats), lease hunting and ecotourism.

8. CONCLUSIONS AND MANAGEMENT IMPLICATIONS

Early successional models of rangeland degradation probably encouraged overly optimistic attitudes towards rangeland management, especially with respect to the potential for rangelands to recover from excessive utilization. However, for the past decade, there has been growing acceptance that anthropogenic disturbances often produce altered vegetation states that are stable enough to make the changes practically irreversible (Natural Resources Council, 1994; Rapport and Whitford, 1999). Widespread observations indicate that once critical disturbance thresholds are crossed, degradation of the structure and function of rangelands will not necessarily be halted or reversed simply by decreasing or removing anthropogenic stresses. In rangelands where livestock production is socioeconomically important and where excessive grazing has occurred, recovery following destocking or even removal of livestock may be minimal. In some cases, it will not even arrest degradation if positive feedbacks have been set in motion.

Despite the broad recognition that discontinuous change is a common feature of many rangelands, little progress has been made towards quantifying transition thresholds for management applications. Management should (a) seek to minimize displacement over critical degradation thresholds; (b) recognize that infrequent, extreme environmental events (e.g., drought) may push systems past critical thresholds and (c) explicitly identify process constraints to recovery following relaxation or removal of stress. A major challenge now facing rangeland ecologists is to identify and characterize ecological processes, plant-soil biota population parameters and soil physical/chemical properties that may provide managers advance warning that a critical transition threshold is being approached. Armed with such information, management could then be adjusted to avert undesirable transitions or to initiate strategic actions to promote desirable transitions. Criteria for classifying landscapes with regard to their transition probability would help prioritize allocation of scarce management resources. Some lands may be past a point of no return; others may be slated to minimize further degradation; still others may be candidates for investing resources to remove constraints to recovery and promote restoration. Reversal of transitions may often require active intervention by land managers. In areas experiencing 'desertification' traditional agronomic approaches to ecosystem restoration are management intensive, costly and seldom successful. Current ecological approaches advocate selective intervention on strategic landscape elements to concentrate scarce resources (water, nutrients, propagules) and establish species capable of initiating autogenic succession and driving progressive rehabilitation (Whisenant et al., 1995; Havstad et al., this volume; Ludwig and Tongway, this volume; Thurow, this volume).

ACKNOWLEDGEMENTS

We thank the organizers of the Rangelands Desertification workshop for the opportunity to participate in this endeavor. The manuscript was improved by helpful comments provided by A. Aradottir. Chad McMurtry assisted with graphics and Veronica Macias helped finalize the references. Their help is much appreciated.

REFERENCES

Aradottir, A., Arnalds, O. and Archer, S. 1992. Degradation of vegetation and soils. In: Icelandic Yearbook of Soil Conservation. Ed. A. Arnalds. pp. 73–82. Soil Conservation Service, Gunnarsholt, Iceland.

Archer, S. 1989. Have southern Texas savannas been converted to woodlands in recent history? American Nat. 134, 545–561.

Archer, S. 1994. Woody plant encroachment into southwestern grasslands and savannas: rates, patterns and proximate causes. In: Ecological Implications of Livestock Herbivory in the West. Ed. M. Vavra, Laycock, W. and R. Pieper. pp. 13–68. Soc. Range Management, Denver, CO.

Archer, S. 1995. Tree-grass dynamics in a Prosopis-thornscrub savanna parkland: reconstructing the past and predicting the future. Ecoscience 2, 83–99.

Archer, S. 1996. Assessing and interpreting grass-woody plant dynamics. In: The Ecology and Management of Grazing Systems. Ed. J. Hodgson and A. Illius. pp. 101–134. CAB International, Wallingford, Oxon, UK.

Archer, S., Scifres, C.J., Bassham, C.R. and Maggio R. 1988. Autogenic succession in a subtropical savanna: conversion of grassland to thorn woodland. Ecol. Monogr. 58, 111–127.

Archer, S. and Smeins, F.E. 1991 Ecosystem-level processes. In: Grazing Management: An Ecological Perspective. Ed. R. K. Heitschmidt and J. W. Stuth. pp. 109–139. Timberline Press, Portland, OR.

Arnalds, A. 1987. Ecosystem disturbance in Iceland. Arctic Alpine Res. 19, 508–513.

Arnalds, O. 1990. Characterization and erosion of Andisols in Iceland. Ph.D. thesis, Texas A&M University.

Arnalds, O. 1998. Desertification in Iceland. Desertification Control Bulletin 32, 22–24.

Arnalds O. 1999. The Icelandic rofabard soil erosion features. Earth Surface Processes and Landforms. (In press).

Baisan, C.H. and Swetnam, T.W. 1990. Fire history on a desert mountain range: Rincon Mountain Wilderness, Arizona, USA. Can. J. For. Res. 20, 1559–1569.

Bosch, O.J.H. and Booysen, J. 1992. An integrative approach to rangeland condition and capability assessment. J. Range Manage. 45, 116–122.

Branson, F.A. 1985. Vegetation changes on western rangelands Vol. 2. Soc.Range Management, Denver, CO.

Chapin, F.S., III, Bloom, A.J., Field, C.B. and Waring, R.H. 1987. Plant responses to multiple environmental factors. BioScience 37, 49–57.

Chew, R.M. 1982. Changes in herbaceous and suffrutescent perennials in grazed and ungrazed desertified grassland in southeastern Arizona, 1958–1978. Am. Midl. Nat. 108, 159–169.

Clarkson, N.M. and Lee, G.R. 1988. Effects of grazing and severe drought on a native pasture in the Traprock region of southern Queensland. Trop. Grassl. 22, 176–183.

Collins, S.L. 1987. Interaction of disturbances in tallgrass prairie: a field experiment. Ecology 68, 1243–1250.

Collins, S.L. and Barber, S.C. 1985. Effects of disturbance on diversity in mixed-grass prairie. Vegetatio 64, 87–94.

Connell, J.H. and Slatyer, R.O. 1977. Mechanisms of succession in natural communities and their role in community stability and organization. American Nat. 111, 1119–1144.

D'Antonio, C.M. and Vitousek, P.M. 1992. Biological invasions by exotic grasses, the grass/fire cycle, and global change. Ann. Rev. Ecol. Syst. 23, 63–87.

Dublin, H.T., Sinclair, A.R.E. and McGlade, J. 1990. Elephants and fire as causes of multiple stable states in the Serengeti-Mara woodlands. J. Animal Ecol. 49, 1147–1164.

Dyksterhuis, E.J. 1949 Condition and management of rangeland based on quantitative ecology. J. Range Manage. 2, 104–105.

Flinn, R.C., Scifres, C.J. and Archer, S.R. 1992. Variation in basal sprouting in co-occurring shrubs: implications for stand dynamics. J. Vegetation Sci. 3, 125–128.

Foran, B.D. 1986. The impact of rabbits and cattle on an arid calcareous shrubby grassland in central Australia. Vegetatio 66, 49–59.

Friedel, M.H. 1991. Range condition assessment and the concept of thresholds: A viewpoint. J. Range Manage. 44, 422–426.

Fulbright, J.E. and Beasom, S.L. 1987 Long-term effects of mechanical treatment on white-tailed deer browse. Wildlife Soc. Bull. 15, 560–564.

Gadgill, M. and Meher-Homji, V.M. 1985. Land use and productive potential of Indian savannas. In: Ecology and Management of the World's Savannas. Eds. J.C. Tothill and J.J. Mott. pp. 107–113. Australian Academy of Science, Canberra.

George, M.R., Brown, J. R. and Clawson, W. J. 1992. Application of non-equilibrium ecology to management of Mediterranean grasslands. J. Range Manage. 45, 436–440.

Godran, M. and Forman, R.T.T. 1983. Landscape modification and changing ecological characteristics. In: Disturbance and Ecosystems. Eds. H.A. Mooney and M. Godron. pp. 12–28. Ecological Studies 44. Springer-Verlag, New York.

Herbel, C.H., Ares, F.N. and Wright, R.A. 1972. Drought effects on a semi-desert grassland. Ecology 53, 1084–1093.

Hibbard, K.A. 1995. Landscape patterns of carbon and nitrogen dynamics in a subtropical savanna: observations and models. Ph.D. dissertation, Texas A&M Univ., College Station.

Hobbie, S.E. 1992 . Effects of plant species on nutrient cycling. Trends Ecol. Evol. 7, 336–339.

Hobbs, R.J. 1994. Dynamics of vegetation mosaics: Can we predict responses to global change? Ecoscience 1, 346–356.

Jones, R.M. 1992. Resting from grazing to reverse changes in sown pasture composition: application of the 'state-and-transition' model. Trop. Grassl. 26, 97–99.

Joyce, L.A. 1993. The life cycle of the range condition concept. J. Range Manage. 46, 132–138.

Kieft, T.L., White, C.S., Loftin, S.R., Aguilar, R., Craig J.A. and Skaar D.A. 1998. Temporal dynamics in soil carbon and nitrogen resources at a grassland-shrubland ecotone. Ecology 79, 671–683.

Laycock, W.A. 1991 Stable states and thresholds of range condition on North American rangelands: A viewpoint. J. Range Manage. 44, 427–433.

Lockwood, J.A. and Lockwood, D.R. 1993 Catastrophe theory – a unified paradigm for rangeland ecosystem dynamics. J. Range Manage. 46, 282–288.

Loucks, O.L., Plumb-Mentjes, M.L. and Rogers, D. 1985. Gap processes and large-scale disturbances in sand prairies. In: The Ecology of Natural Disturbance and Patch Dynamics. Eds. S.T.A. Picket and P.S. White. pp. 71–83. Academic Press, Inc., New York.

Luken, J. 1990. Directing Ecological Succession. Chapman and Hall, London.

Mack, R.N. 1981. Invasion of Bromus tectorum L. into western North America, and ecological chronicle. Agroecosystems 7, 145–165.

McNaughton, S.J. 1983. Serengeti grassland ecology: the role of composite environmental factors and contingency in community organization. Ecol. Monogr. 53, 291–320.

Milton, S.J. 1992. Effects of rainfall, competition and grazing on flowering of Osteospermum sinuatum (Asteraceae) in arid Karoo rangeland. J. Grassland Soc. South Africa 9, 158–164.

Milton, S.J. and Hoffman, M.T. 1994 The application of state-and-transition models to rangeland research and management in arid succulent and semi-arid grassy Karoo, South Africa. African J. Range Science 11, 18–26.

National Research Council 1994. Rangeland health: new methods to classify, inventory and monitor rangelands National Academy Press, Washington, DC.

Neilson, R.P. 1986 High-resolution climatic analysis and Southwest biogeography. Science 232, 27– 34.

O'Connor, T.G. 1993. The influence of rainfall and grazing on the demography of some African savanna grasses: a matrix modelling approach. J. Appl. Ecol. 30, 119–132.

O'Connor, T.G. 1994. Composition and population responses of an African savanna grassland to rainfall and grazing. J. Appl. Ecol. 31, 155–171.

Orr, D.M., Evenson, C.J., Lehane, J.K., Bowly, P.S. and Cowan, D.C. 1993. Dynamic of perennial grasses with sheep grazing in Acacia aneura woodlands in south-west Queensland. Trop. Grassl. 27, 87–93.

Overpeck, T., Rind, D. and Goldberg, R. 1990. Climate-induced changes in forest disturbance and vegetation. Nature 343, 51–53.

Rapport, D.J. and Whitford, W.G. 1999. How ecosystems respond to stress. BioScience 49, 193–203.

Reid, N., Marroquín, J., Beyer-Münzel, P. 1990. Utilization of shrubs and trees for browse, fuelwood and timber in Tamaulipan thornscrub, northeastern Mexico. Forest Ecol. Manage. 36, 61–79.

Rykiel, E.J. Jr. 1985. Towards a definition of ecological disturbance. Australian J. Ecology 10, 361–365.

Savage, M. and Swetnam, T.W. 1990. Early 19th-century fire decline following sheep pasturing in a Navajo ponderosa pine forest. Ecology 71, 2374–2378.

Schlatterer, E.F. 1989. Toward a user-friendly ecosystem: myth or mirth? In: Proceedings, Land Classifications, Based on Vegetation: Applications for Resource Management. Eds. D.E. Ferguson, P. Morgan and F.D. Johnson. pp. 223–227. General Technical Report INT-257, USDA Forest Service, Intermountain Research Station, Ogden, Utah.

Schlesinger, W.H., Reynolds, J.F., Cunningham, G.L., Huenneke, LF., Jarrell, W.M., Virginia, R.A. and Whitford, W.G. 1990. Biological feedbacks in global desertification. Science 247, 1043–1048.

Scholes, R. and Archer, S. 1997. Tree-grass interactions in savannas. Ann. Rev. Ecol. Syst. 28, 517–544.

Scifres, C.J., Hamilton, W.T., Inglis, J.M. and Conner, J.R. 1983. Development of integrated brush management systems (IBMS): decision-making processes. In: Proceedings Brush Management Symposium. Ed. K.W. McDaniel. pp. 97–104. Texas Tech Press, Lubbock.

Smeins, F.E. and Merrill, L.B. 1988. Long-term change in semi-arid grassland. In: Edwards Plateau Vegetation. Eds. B.B. Amos and F.R. Gehlback. pp. 101–114. Baylor Univ. Press, Waco, TX.

Stokes, C.J. 1994 Degradation and Dynamics of Succulent Karoo Vegetation. Published, Place University of Natal, Pietermaritzburg, South Africa.

Swetnam, T.W. and Betancourt, J.L. 1990. Fire-southern oscillation relations in the southwestern United States. Science 249, 1010–1020.

Verstraete, M.M. 1986. Defining desertification: a review. Climatic Change 9, 5–18.

Watson, I.W., Burnside, D.G. and Holm, A.M. 1996. Event-driven or continuous: which is the better model for managers? The Rangeland J. 18, 351–369.

Watson, I.W., Westoby, M. and Holm A.M. 1997a. Continuous and episodic components of demographic change in arid zone shrubs: models of two Eremophila species from western Australia compared with published data on other species. J. Ecology 85, 833–846.

Watson, I.W., Westoby, M. and Holm, A.M. 1997b. Demography of two shrub species from an arid grazed ecosystem in Western Australia 1983–93. J. Ecology 85,815–832.

West, N. E . 1986. Desertification or xerification? Nature 321, 562–563.

Westoby, M., Walker, B. and Noy-Meir, I. 1989a. Opportunistic management for rangelands not at equilibrium. J. Range Manage. 42, 266–274.

Westoby, M., Walker, B. and Noy-Meir, I. 1989b. Range management on the basis of a model which does not seek to establish equilibrium. J. Arid Environ. 17, 235–239.

Whisenant, S.G., Thurow, T.T. and Maranz, S.J. 1995. Initiating autogenic restoration on shallow semiarid sites. Restoration Ecology 3, 61–67.

Whitford, W.G., Martinez-Turanzas, G. and Martinez-Meza, E. 1995. Persistence of desertified ecosystems: explanations and implications. In: Desertification in Developed Countries. Eds. D.A. Moat and C.F. Hutchinson. pp. 319–332. Kluwer Academic Publishers, London.

Yassoglou, N.J. and Kosmas, C. 1999. Desertification in the Mediterranean Europe. In: Rangeland Desertification, International Workshop, Iceland, September 1997. Eds. O. Arnalds and S. Archer. RALA Report no. 200. (In press).

Viewing rangelands as landscape systems

John A. Ludwig[1] and David J. Tongway[2],
[1] *CSIRO Wildlife and Ecology, PMB 44, Winnellie, Darwin, NT, 0822 Australia*
[2] *CSIRO Wildlife and Ecology, P.O. Box 84, Lyneham, Canberra, ACT, 2602 Australia*
Tel: 618 8944 8423, Fax: 618 8944 8444; E-mail: <john.ludwig@terc.csiro.au>

ABSTRACT In this paper we take a fresh look at how to reach a predictable understanding of how rangelands work by viewing them as landscape systems[3]. What are the landscape structures and processes operating in rangelands that allow us to better understand how limited resources, such as water and soil nutrients, are naturally conserved within these systems? How can this understanding help us to better manage the use of rangelands so that degradation can be prevented? Can we distinguish between utilization and desertification? These kinds of questions can be addressed within a logical framework based on landscape ecology. This framework, labeled the trigger-transfer-reserve-pulse (TTRP) conceptual model, is built upon existing concepts applied to arid and semi-arid, resource-limited rangelands. The TTRP framework focuses on how materials in short-supply such as water and soil nutrients (in sediments) are carried by runoff, or by winds, and are captured and concentrated within landscape patches (reserves). If the availability of these resources within a patch exceeds a critical threshold, a production 'pulse' occurs. The organic structures and materials produced in such pulses feedback to build on the integrity of landscape patches, and through them to the landscape as a whole, enabling them to capture more resources in the future. However, if patches are degraded by disturbances such as grazing, fewer scarce resources will be captured and landscapes can become dysfunctional. The relative 'state' of dysfunction for a rangeland unit can be measured and used to position the unit along a continuum of landscape functionality, from highly functional to highly dysfunctional. Whether this position on the continuum is acceptable for the current uses being made of the rangeland can then be evaluated. If an undesirable change in landscape functionality is detected by monitoring indicators on a unit of rangeland, then management recommendations and strategies can be developed to reverse this change by viewing the rangeland unit as a dynamic landscape system functioning to conserve resources.

Key words: landscape ecology, landscape function, landscape patches, rangeland degradation, runoff, soil nutrients.

---

[3] This paper summarizes some of the concepts and principles described in more detail in various Chapters of "Landscape Ecology, Function and Management: Principles from Australia's Rangelands" (Ludwig et al., 1997).

1. INTRODUCTION

Range scientists and practitioners have built a fairly comprehensive understanding of rangeland ecosystems over the years (e.g., Committee on Developing Strategies for Rangeland Management, 1984). However, it is often useful to step back and take a fresh look at this understanding. In his address to the Fifth Rangeland Congress, Walker (1996) called for an understanding of how rangelands function by building conceptual models. In this paper, after providing some definitions, we present a conceptual framework for describing how rangelands function as landscape systems to conserve scarce resources – a framework based on the principles of landscape ecology and function (Ludwig et al., 1997). We have found this framework useful for assessing the impacts of different rangeland uses (Tongway and Ludwig, 1993). We suggest that by regarding rangelands as landscape systems, their responses to different uses can be predicted by using simulation models (Ludwig and Marsden, 1995). Management strategies can then be developed to help us use rangelands in a sustainable manner (Ludwig and Freudenberger, 1997).

1.1 Background definitions

In landscape ecology, the basic unit is defined as two or more units of land, functionally linked as a network by their source-sink relationships (Risser, 1987). For example, if water and sediments are running off one range unit onto an adjacent unit, then these two units can be considered as one landscape unit or system. The sizes of these units will depend on research or management objectives. Much of the literature on landscape ecology deals with large areas, such as regions of 100 km^2 or more (e.g., Turner and Gardner, 1991). However, our research objectives are usually aimed at understanding those landscape processes which operate to conserve scarce resources such as water and soil nutrients within smaller landscape units, usually of less than 10 ha (e.g., Tongway and Ludwig, 1997). At this local landscape system size, processes operate at scales of less than 10 m^2 (see also Havstad et al., this volume). We believe that these are the scales at which one must manage rangelands for grazing impacts (Tongway and Ludwig, 1993) or for biodiversity (Freudenberger et al., 1997; for 'watershed' perspective see Thurow, this volume).

Local landscape units can also have a distinctive structural organization (Ludwig and Tongway, 1995). For example, in the semi-arid woodlands of eastern Australia, tree groves and intergroves form toposequences. In other words, a tree grove-intergrove landscape unit will repeat itself down a local topographic gradient, that is, from the top of a ridge down to the bottom of a swale. These landscape toposequence units function to capture resources. For example, water and nutrients lost from an upslope tree grove may be captured by the grove below it. At a very fine-scale, or micro-scale of less than 1 m^2, are very small landscape patches such as grass tussocks which also function to capture water and nutrients (Tongway and Ludwig, 1994). Important biological processes occur at this scale, for example, litter is processed by termites within such grass patches (e.g., Whitford et al., 1992).

2. A LANDSCAPE FRAMEWORK FOR RANGELAND SYSTEMS

A framework is a structure to build upon. Below, we describe how rangelands function as landscape systems built upon four components: trigger, transfer, reserve and pulse (Fig. 1). Arrows between components indicate the direction of flows of resource materials such as water, soil sediments and nutrients, litter and seeds.

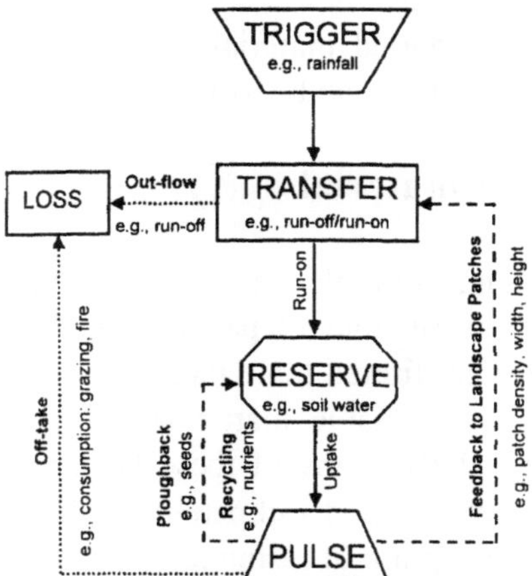

Figure 1. The trigger-transfer-reserve-pulse (TTRP) framework for arid and semi-arid rangelands. (After Fig. 1.1; Ludwig and Tongway, 1997).

2.1 The trigger

In rangelands, and most other systems, rainfall is the primary trigger (Fig. 1); it initiates biological, physical and chemical events (Noy-Meir, 1973). Others emphasize that nutrients (e.g., soil nitrogen) are critical secondary regulators of arid and semi-arid systems (Whitford et al., 1987; Havstad et al., this volume). For example, field experiments have demonstrated the importance of soil nutrients in regulating the sizes of growth pulses when rainfall or irrigation makes soil water available (Gutierrez et al., 1988). Rainfall triggers are highly unpredictable in amount, season, and distribution in most arid and semi-arid rangelands. Thus, when they occur, they trigger important events.

2.2 Transfer of materials

As used in the TTRP framework (Fig. 1), the horizontal redistribution of resource materials across a landscape by water or wind is a transfer. For example, rainwater is transferred into tree groves by runoff-runon (Greene et al., 1994) which also transfers soil sediments and nutrients. Runoff is critically important in the arid and semi-arid rangelands of Australia because their ancient soils tend to have very low infiltration rates. Greene et al. found infiltration rates of less than 10 mm/h, and when they experimentally applied water at 30 mm/h, runoff started in a few minutes and carried sediment loads of over 0.3 g/l. Wind also transfers soil and litter particles in rangeland landscapes. Unless these transported materials are captured by the landscape, they will flow out of the system and be lost.

2.3 Reserves for materials

Materials carried along in runoff water, or by winds, can be held up and captured by landscape obstructions or patches (reserves). Field and laboratory measurements have confirmed that landscape patches capture resources such as soil nutrients and organic matter (Ludwig and Tongway, 1995; Tongway and Ludwig, 1990). Many of these landscapes are distinctly patterned, with discrete vegetation and soil patches at different scales; and these patterns are linked to the capture of resources by patches through both water and wind-mediated processes (Ludwig and Tongway, 1992; Tongway, 1993). Thus, patches within rangeland systems, from small grass tussocks to large tree groves, function as reserves for soil water and nutrients.

2.4 Pulse events and flows

If a big rainfall event or trigger occurs, and if sufficient runoff is captured by landscape patches or reserves, then a pulse of growth will occur within these patches (Fig. 1). Of course, each organism (plant, animal or microbial) living in the patch, will have its own critical threshold for having a pulse, depending on how it has evolved to survive in rangeland environments. For example, trees may produce a pulse or flush of new leaves – the old leaves having been dropped in the dry period prior to the rainfall trigger. If their growth pulse is large, these trees may produce flowers, fruits and seeds. These seeds, if dropped in the patch, represent a 'plough back' of resources into the landscape patch or reserve, rebuilding seed pools or banks in the litter or soil.

If a rainfall trigger causes a growth pulse through seed germination and establishment, these new plants represent a 'feedback' to the landscape patch (Fig. 1). For example, if new grass tussocks grow within a patch, thereby increasing the total density of tussocks in the patch, this will increase the ability of the patch to capture or trap runoff, sediments and litter when the next rainfall trigger occurs. Of course, some of the growth produced by plants may be eaten by animals which leave the landscape system (e.g., grasshoppers fly away, sheep are mustered and sold); this represents 'off-take' from the system. Fires can burn the landscape, with smoke and ash blowing out of the system. Back at the time of the rainfall event, if the rate of runoff exceeds the capacity of the landscape patches to trap and store this water, then 'out-flow' occurs. These flows, out of the landscape system of interest, may be down creeks or rivers, or into lakes or salt pan.

3. MANAGING RANGELANDS AS PATCHY LANDSCAPE SYSTEMS

As described above and documented elsewhere (Tongway and Ludwig, 1997a), many arid or semi-arid rangelands are patchy systems, which trap and conserve scarce resources. The TTRP framework provides a conceptual structure for the physical and biological processes involved (e.g., runoff, runon, plough backs, feedbacks). Why is an understanding of landscape patch structures and processes important for managing rangelands?

3.1 Reversing the 'Robin Hood' affect

From the English ballads of the late 14th and early 15th century, Robin Hood is a character who was portrayed as a hero (at least for the poor) because he 'robbed the rich to give to the poor'. However, in many highly resource-limited rangelands with very unpredictable rainfall, this principle attributed to Robin Hood is reversed, that is, 'poor [landscape zones] are robbed to pay the rich' (Tongway and Ludwig, 1997a). For example, when landscape patches are enriched by resources flowing in runoff from inter-patch areas, this represents a transfer of resources from poorer areas in the landscape (runoff or source zones) to rich areas (runon or sink zones). In other words, already poor runoff areas are being robbed of their meager resources for the benefit of the already resource-rich runon patches. Is this a natural phenomenon, and if so, why?

Noy-Meir (1981) argued that if rainwater from a moderately low rainfall event (a trigger) is spread evenly or uniformly over the surface of an arid ecosystem (i.e., no runoff or transfer), then not enough water would be stored in the soil to meet the threshold requirements for biological activity (a pulse). He derived mathematical models to predict that if water is transferred by runoff processes and concentrated into patches in these landscapes, then there would be adequate soil water for pulses of growth (at least within these patches). Thus, the total productivity of patchy landscapes in resource-limited rangelands is greater than for non-patchy rangelands in similar environments. Simulation studies, using a mathematical model structured on the TTRP framework, and validated with production data from a semi-arid rangeland, confirm this prediction (Ludwig et al., 1994). In fact, this predicted phenomenon has long been applied as a 'water harvesting' practice in some arid lands to enhance plant production (e.g., Dodd and Skinner, 1990). Rangelands need to be managed carefully to maintain their natural patchiness if forage and animal production are to be maintained in resource-poor arid and semi-arid landscape systems.

3.2 Maintaining a balanced system

Another way to view the importance of maintaining patchiness in rangelands is to compare landscape systems that are 'in balance' with those that are 'out of balance' (Fig. 2). Using a simplified patch-pulse version of the TTRP framework, a rangeland that is being managed well can be viewed as a landscape system where there is a good balance between feedbacks and off-take losses (Fig. 2a). In other words, the landscape system is balanced because patches are capturing water and nutrients, producing good growth pulses, and then gaining feedbacks from these pulses while losing some production to consumption by animals (or fire).

However, if an excessive amount of the water and nutrients is lost from the landscape system, or if a production pulse does occur but most of this pulse is consumed and taken off the landscape, then the system is out of balance (Fig. 2b). In other words, if runoff is not being efficiently captured and stored within landscape patches (system reserves), production pulses are likely to be small and, in grazed rangelands, what is produced is likely to be consumed and removed. With smaller feedbacks to the patches, their structural integrity will not be maintained (e.g., fewer and weaker grass tussocks). If this continues over a long time period, patches are likely to degenerate and the system will have a low capacity for capturing runoff (Anderson and Hodgkinson, 1997). These degraded landscapes have been called dysfunctional (Tongway and Ludwig, 1997b).

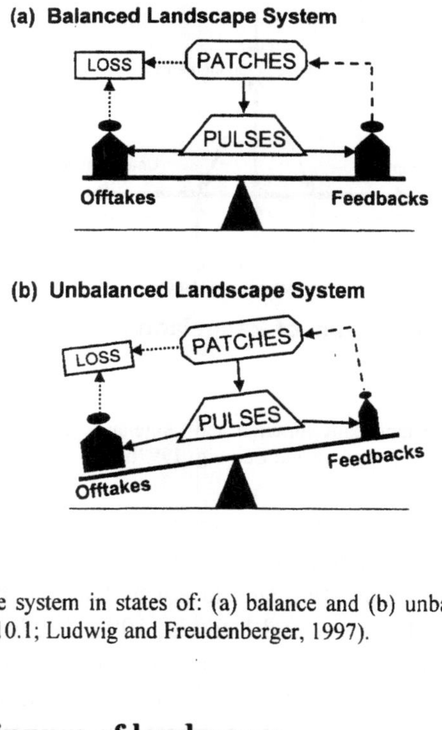

Figure 2. A Landscape system in states of: (a) balance and (b) unbalance in off-takes and feedbacks. (After Fig. 10.1; Ludwig and Freudenberger, 1997).

3.3 A continuum of landscapes

Maintaining the natural patchiness of rangelands, hence keeping them as balanced landscape systems, is a worthwhile goal of management, both ecologically and economically. Of course, many rangelands are not naturally patchy, except perhaps at such a fine scale (e.g., 5–10 cm) that they appear to be uniform (e.g., grasslands on fertile loam and clay soils in higher rainfall zones). However, for those rangelands that are distinctly patchy landscape systems, this patchiness can vary greatly depending on how they are managed. Such landscapes can vary along a continuum from those that have 'healthy' patches which conserve water and nutrient resources to those having degraded patches which are 'leaky' (i.e., lose scarce resources) (Fig. 3). These landscapes can occur along a continuum of how well they function, from fully functional resource conserving systems to totally dysfunctional, leaky systems. A totally dysfunctional landscape will have no patches, only open, bare space, which completely sheds resources. A fully functional landscape will have enough patches to completely capture all resources.

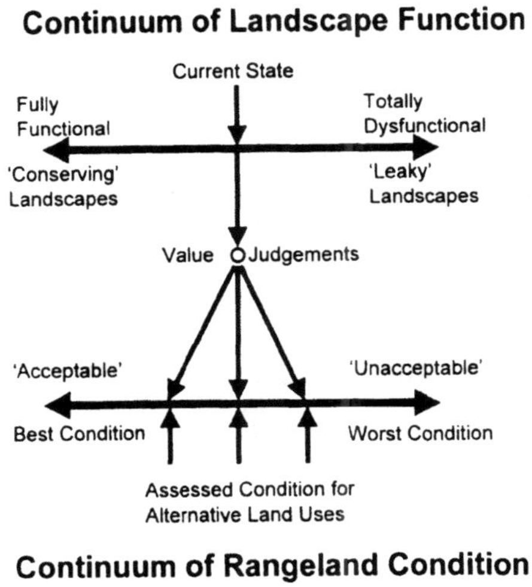

Figure 3. Continuums for landscape function and rangeland condition, linked through value judgements. (After Fig. 5.1; Tongway and Ludwig, 1997b).

The position of a landscape along this functionality continuum only provides an indication of the potential ability of a given landscape to trap, store and utilize resources. Its position should not be taken as being equivalent to rangeland condition (Tongway and Ludwig, 1995, 1997b). It is very important to distinguish between a continuum of landscape functionality and a continuum of rangeland condition (Fig. 3). This is because whether resource losses from a landscape are excessive or not, or whether the level of patch degradation is acceptable or unacceptable, is very much a personal value judgement; such judgements depend on one's purpose for using the land, and on one's social views and cultural upbringing (Young et al., 1984). For example, a landscape whose state of patchiness, hence function, is intermediate, may be judged as being in good condition if it is being used for wool production and if this production has not been significantly impacted by a loss of patches. This is the case for many of the chenopod shrublands where sheep grazing has reduced distinctive shrub patches, but a replacement by ephemeral grasses has maintained wool production (Graetz, 1986). However, a park ranger may judge this same shrubland with intermediate landscape functionality as a landscape system in poor condition because it does not provide the habitat patches (shrubs) needed to protect an endangered animal.

4. INDICATORS OF LANDSCAPE FUNCTIONALITY

If rangeland has been degraded in terms of landscape function, how do we determine where it is along the function-dysfunction continuum? Are there any indicators of landscape function that are easy to measure?

4.1 Patchiness and soil surface indicators

Four measures of landscape patchiness have proven to be good indicators of landscape function (Tongway and Ludwig, 1997b; Tongway and Hindley, this volume): (1) the number of patches by type, as counted along a line transect oriented downslope; (2) the length of each patch intercepting the line, which estimates total cover of patches; (3) the width of each patch, as measured perpendicular to the line; and (4) the fetch length between patches, which estimates the average distance from patch to patch along the line. When these four indicators are measured on rangelands judged to be in good and poor range condition, a comparison is instructive. The landscape in poor range condition will have few patches, and these will be of low total cover, narrow in average width, and they will have long distances between them; the converse will be found for landscapes in good range condition (Tongway and Ludwig, 1997b). In wind-driven landscapes, transects can be oriented downwind and measures such as shrub canopy horizontal-cover used to indicate landscape function and rangeland condition (Andre Van Rooyan, pers. com.).

The condition of the soil surface is another sensitive indicator which can be measured along a line transect. Soil surface condition measures indicate a number of landscape attributes which are difficult to measure directly, such as infiltration and erosion rates, and how well nutrient cycles are working. If the line transect is permanently marked, then landscape and soil surface indicators can be monitored through time. This can provide a rangeland manager with valuable information about whether the landscape is becoming more degraded (e.g., less patchy), or recovering its ability to capture resources (e.g., more patchy). Instructions on how to precisely measure landscape patchiness and soil surface condition indicators are provided in Tongway (1994) and Tongway and Hindley (1995).

4.2 Early warning indicators

Australia has signed the UN Convention on Desertification (Williams et al., 1995), pledging to prevent the degradation of lands vulnerable to desertification. To prevent rangeland degradation, one must have early warning signs or indicators that signal when a landscape is becoming desertified or dysfunctional. Landscape patchiness and soil surface condition indicators, if carefully monitored, should provide an early warning signal that landscapes are moving along the continuum in the direction of dysfunction (i.e., toward becoming desertified). Other indicators, such as the 'health' of perennial grass populations may also serve as early warning signs (e.g., Anderson and Hodgkinson, 1997). However, further research is needed on whether these indicators are sensitive enough to signal an early warning of desertification.

5. RESTORING RANGELANDS AS PATCHY LANDSCAPE SYSTEMS

If a landscape has already been degraded or desertified (i.e., become partially dysfunctional), can it regain its ability to capture resources by simply removing the cause of the degradation (Milton et al., 1994; Archer and Stokes, this volume; Hoffman, this volume)? Are patches of vegetation and soil surfaces still sufficiently intact such that reducing grazing off-takes will provide adequate feedbacks to patches to restore the capacity of the landscape to trap scarce resources? Range scientists debate whether degraded rangelands can recover by themselves (e.g., Ludwig et al., 1996). The time required for rangelands to 'heal' is often central to this debate. For example, a degraded rangeland located in eastern Australia was intensively studied in the 1970s (e.g., Harrington et al., 1981). An exclosure, proofed against sheep, goat, kangaroo and rabbit grazing, took 20 years to show signs of recovery (Tongway and Ludwig, 1997b). If this is too long to wait, then rehabilitation is needed (e.g., Ludwig and Tongway, 1996; Tongway and Ludwig, 1996).

6. SUSTAINING RANGELANDS AS LANDSCAPE SYSTEMS

How are rangeland managers going to maintain functional landscapes into the future when faced with problems such as higher costs, lower prices, global competition, climate change, cultural change, and an increasing demands for using rangelands for purposes other than pastoralism (Foran et al., 1990; Ludwig and Freudenberger, 1997)? What level of landscape functionality will be acceptable to all the different users of rangelands, each with their own sets of values (Holmes and Day, 1995; MacLeod and Taylor,

1994)? How can rangeland functionality be sustained in the face of all these demands? Are land use optimization procedures needed (Walker, 1996)?

Perhaps the key is to understand landscape structures and processes, and how these regulate the conservation of scarce resources. In resource-limited, naturally patched rangelands, maintaining landscape patchiness and soil surface condition is how to conserve resources. This understanding is essential if one is to develop effective rangeland management strategies in the face of different land uses and goals. For example, if a rangeland manager wants to maintain biodiversity, then a management strategy to conserve a great variety of patchy habitats must be developed. However, if the rangeland management goal is to produce a consistent supply of wool, then a strategy to maintain perennial grasses as long-lasting, nutritious forage must be developed. All the different users of rangelands will need to participate in the development of such management strategies (Foran et al., 1990; A. Arnalds, this volume).

ACKNOWLEDGEMENTS

We acknowledge our CSIRO colleagues who contributed to the book upon which this paper is based, and many other colleagues in Australia and overseas who helped us formalize and 'solidify' our thinking on landscape function and its application to rangelands. We thank two reviewers for suggestions on ways to improve a draft of this paper.

LITERATURE CITED

Anderson, V.J. and Hodgkinson, K.C. 1997. Grass-mediated capture of resource flows and the maintenance of banded mulga in a semi-arid woodland. Australian Journal of Botany 45, 331–342.

Committee on Developing Strategies for Rangeland Management. 1984. Developing Strategies for Rangeland Management. A Report Prepared by the Committee on Developing Strategies for Rangeland Management for the U.S. National Research Council and the National Academy of Sciences. Westview Press, Boulder, 2022 pp.

Dodd, J.L. and Skinner, Q.D. 1990. Water harvesting and management in arid lands. In: Advances in Range Management in Arid Lands. Eds. R. Halwagy, F.K. Taha and S.A. Omar. pp. 101–112. Kegan Paul International, London.

Foran, B.D., Friedel, M., MacLeod, N.D., Stafford Smith, D.M. and Wilson, A.D. 1990. Policy Proposals for the Future of Australia's Rangelands. Australian Government Publishing Service, Canberra.

Freudenberger, D., Noble, J. and Hodgkinson, K. 1997. Management for production and conservation goals in rangelands. Chapter 8. In: Landscape Ecology, Function and Management: Principles from Australia's Rangelands. Eds. J. Ludwig, D. Tongway, D. Freudenberger, J. Noble and K. Hodgkinson. pp. 93–106. CSIRO Publishing, Melbourne.

Graetz, R.D. 1986. A comparative study of sheep grazing in a semi-arid saltbush pasture in two condition classes. Australian Rangeland Journal 8, 46–56.

Greene, R.S.B., Kinnell, P.I.A. and Wood, J.T. 1994. Role of plant cover and stock trampling on runoff and soil erosion from semi-arid wooded rangelands. Australian Journal of Soil Research 32, 953–973.

Gutierrez, J.R., DaSilva, O.A., Pagani, M.I., Weems, D. and Whitford, W.G. 1988. Effects of different patterns of supplemental water and nitrogen fertilization on productivity and composition of Chihuahuan Desert annual plants. American Midland Naturalist 119, 336–343.

Harrington, G.N., Dawes, G.T. and Ludwig, J.A. 1981. An analysis of the vegetation pattern in a semi-arid *Eucalyptus populnea* woodland in north-west New South Wales. Australian Journal of Ecology 6, 279–287.

Holmes, J.H. and Day, P. 1995. Identity, lifestyle and survival: value orientations of South Australian pastoralists. Rangeland Journal 17, 193–212.

Ludwig, J.A. and Freudenberger, D. 1997. Towards a sustainable future for rangelands. Chapter 10. In: Landscape Ecology, Function and Management: Principles from Australia's Rangelands. Eds. J. Ludwig, D. Tongway, D. Freudenberger, J. Noble and K. Hodgkinson. pp. 121–131. CSIRO Publishing, Melbourne.

Ludwig, J.A. and Marsden, S.G. 1995. A simulation of resource dynamics within degraded semi-arid landscapes. Mathematics and Computers in Simulation 39, 219–224.

Ludwig, J.A. and Tongway, D.J. 1992. Monitoring the condition of Australian arid lands: linked plant-soil indicators. In: Ecological Indicators, Vol. 1. Eds. D.H. McKenzie, D.E. Hyatt and V.J. McDonald. pp. 765–772 Elsevier, Essex.

Ludwig, J.A. and Tongway, D.J. 1995. Spatial organisation of landscapes and its function in semi-arid woodlands, Australia. Landscape Ecology 10, 51–63.

Ludwig, J.A. and Tongway, D.J. 1996. Rehabilitation of semiarid landscapes in Australia. II. Restoring vegetation patches. Restoration Ecology 4, 398–406.

Ludwig, J.A. and Tongway, D.J. 1997. A landscape approach to rangeland ecology. Chapter 1. In: Landscape Ecology, Function and Management: Principles from Australia's Rangelands. Eds. J. Ludwig, D. Tongway, D. Freudenberger, J. Noble and K. Hodgkinson. pp. 1–12. CSIRO Publishing, Melbourne.

Ludwig, J.A., Fargher, J.D., Foran, B.D., James, C.D., MacLeod, N.D. and McIntyre, S. 1996. Restoration of our earth's rangelands: emergency damage control or faith in self-healing. In: Rangelands in a Sustainable Biosphere. Vol II. Proceedings of the Fifth International Rangeland Congress. Ed. N. West. pp. 65–71. Society of Range Management, Denver.

Ludwig, J., Tongway, D., Freudenberger, D., Noble, J. and Hodgkinson, K. (Eds.). 1997. Landscape Ecology, Function and Management: Principles from Australia's Rangelands. CSIRO Publishing, Melbourne.

Ludwig, J.A., Tongway, D.J. and Marsden, S.G. 1994. A flow-filter model for simulating the conservation of limited resources in spatially heterogeneous, semi-arid landscapes. Pacific Conservation Biology 1, 209–213.

MacLeod, N.D. and Taylor, J.A. 1994. Perceptions of beef cattle producers and scientists relating to sustainable land use issues and their implications for technology transfer. Rangeland Journal 16, 238–253.

Milton, S.J., Dean, W.R.J., du Plessis, M.A. and Siegfried, W.R. 1994. A conceptual model of arid rangeland degradation: the escalating cost of declining productivity. BioScience 44, 70–76.

Noy-Meir, I. 1973. Desert ecosystems: environment and producers. Annual Review of Ecology and Systematics 4, 25–51.

Noy-Meir, I. 1981. Spatial effects in modelling of arid ecosystems. In: Arid-land Ecosystems: Structure, Functioning and Management, Vol. 2. Eds. D.W. Goodall and R.A. Perry. pp. 411–432. Cambridge University Press, Sydney.

Risser, P.G. 1987. Landscape ecology: state of the art. In: Landscape Heterogeneity and Disturbance. Ed. M.G. Turner. pp. 3–14. Springer-Verlag, New York.

Tongway, D.J. 1993. Functional analysis of degraded rangelands as a means of defining appropriate restoration techniques. In: Proceedings of the Fourth International Rangeland Congress. Eds. A. Gaston, M. Kernick and H. LeHouerou. pp. 166–168. Service Central d'Information Scientifique et Technique, Montpellier.

Tongway, D.J. 1994. Rangeland Soil Condition Assessment Manual. CSIRO Publishing, Melbourne.

Tongway, D.J. and Hindley, N. 1995. Manual for Soil Condition Assessment of Tropical Grasslands. CSIRO Wildlife and Ecology, Canberra.

Tongway, D.J. and Ludwig, J.A. 1990. Vegetation and soil patterning in semi-arid mulga lands of Eastern Australia. Australian Journal of Ecology 15, 23–34.

Tongway, D.J. and Ludwig, J.A. 1993. Rehabilitation of minesites and pastoral lands: the ecosystem function approach. In: Proceedings, Goldfields International Conference on Arid Landcare. pp. 51–57. Milean di Russo Pty Ltd., Hamilton Hill.

Tongway, D.J. and Ludwig, J.A. 1994. Small-scale resource heterogeneity in semi-arid landscapes. Pacific Conservation Biology 1, 201–208.

Tongway, D.J. and Ludwig, J.A. 1995. Function and dysfunction in mulga woodlands. In: Ecological Research and Management in the Mulgalands. Eds. M.J. Page and T.S. Beutel. pp. 85–89. University of Queensland Gatton College, Lawes.

Tongway, D.J. and Ludwig, J.A. 1996. Rehabilitation of semiarid landscapes in Australia. II. Restoring productive soil patches. Restoration Ecology 4, 388–397.

Tongway, D.J. and Ludwig, J.A. 1997a. The conservation of water and nutrients within landscapes. Chapter 2. In: Landscape Ecology, Function and Management: Principles from Australia's Rangelands. Eds. J. Ludwig, D. Tongway, D. Freudenberger, J. Noble and K. Hodgkinson. pp. 13–22. CSIRO Publishing, Melbourne.

Tongway, D.J. and Ludwig, J.A. 1997b. The nature of landscape dysfunction in rangelands. Chapter 5. In: Landscape Ecology, Function and Management: Principles from Australia's Rangelands. Eds. J. Ludwig, D. Tongway, D. Freudenberger, J. Noble and K. Hodgkinson. pp. 49–61. CSIRO Publishing, Melbourne.

Turner, M.G. and Gardner, R.H. (Eds.) 1991. Quantitative Methods in Landscape Ecology – The Analysis and Interpretation of Landscape Heterogeneity. Springer-Verlag, New York.

Walker, B.H. 1996. Having or eating the rangeland cake: a developed world perspective on future options. In: Rangelands in a Sustainable Biosphere. Vol II. Proceedings of the Fifth International Rangeland Congress. Ed. N. West. pp. 22–28. Society of Range Management, Denver.

Whitford, W.G., Ludwig, J.A. and Noble, J.C. 1992. The importance of subterranean termites in semi-arid ecosystems in south-eastern Australia. Journal of Arid Environments 22, 87–91.

Whitford, W.G., Reynolds, J.F. and Cunningham, G.L. 1987. How desertification affects nitrogen limitation of primary production on Chihuahuan Desert watersheds. In: Strategies for Classification and Management of Native Vegetation for Food Production in Arid Zones. Eds. E.F. Aldon, C.E. Gonzales Vicente and W.H. Moir. pp. 143–153. General Technical Report RM-150. U.S. Department of Agriculture, Forest Service, Rocky Mountain Forest and Range Experiment Station, Fort Collins.

Williams, M., McCarthy, M. and Pickup, G. 1995. Desertification, drought and landcare: Australia's role in an international convention to combat desertification. Australian Geographer 26, 23–32.

Young, M.D., Gibbs, M., Holmes, W.E. and Mills, D.M.D. 1984. Socio-economic influences on pastoral management. In: Management of Australia's Rangelands. Eds. G.N. Harrington, A.D. Wilson and M.D. Young. pp. 79–93. CSIRO Publishing, Melbourne.

Hydrologic effects on rangeland degradation and restoration processes

Thomas L. Thurow
Rangeland Ecology and Management Department, Texas A&M University, College Station, Texas 77843-2126, USA
Tel: 409 845 3765; Fax: 409 845 6430, E-mail: <t-thurow@tamu.edu>

ABSTRACT Hydrologic attributes which influence ecosystem structure and function include infiltration rate, soil moisture storage capability, precipitation characteristics and rain use efficiency. The degree to which these hydrologic attributes are affected by land use determines the impact of the land use on the ecosystem. Simply removing the initial cause of degradation may not restore the site's production potential, and may not even break the pattern of decline if a self-sustaining cycle of deterioration has developed. Reasons why severely disturbed landscapes may not recover production potential include loss of species, altered species interactions, physical degradation of hydrologic characteristics or nutrient depletion. The economic realities associated with rangeland ecosystem management restrict implementation of widespread restoration strategies requiring intensive management. Therefore both ecological and economic challenges must be solved if restoration is to proceed. A watershed is a natural scale of resolution which provides a logical context for addressing the ecological and economic challenges of restoration. The drainage patterns of a watershed form the framework of energy and nutrient flow that must be understood to address the ecological considerations of management strategies. These flow patterns also provide the context for a more complete (and more favorable) socioeconomic accounting of the serial benefits and costs of investments in rangeland management.

Key words: infiltration, rain use efficiency, restoration, soil structure, watershed.

1. INTRODUCTION

Leonardo da Vinci's observation that "water is the driving force of nature" underscores the fact that water is the essential medium of biogeochemical cycles and of life itself (Smith, 1974). Land use patterns that alter the hydrologic characteristics of a site, therefore, are likely to have broad impact on ecosystem structure and function.

The interrelated mix of water/soil/vegetation characteristics may be manifest in a variety of theoretically stable configurations on any particular site (Rietkerk and Vandekoppel, 1997). Land use patterns, such as fire suppression or increased grazing intensity, may alter the characteristics of the water/soil/vegetation linkages of the current ecosystem, causing a shift across a "threshold" from one stable ecosystem configuration to another (Holling, 1973; May, 1977; Archer and Stokes, this volume). Usually the new ecosystem configuration that forms in response to human-induced changes in land use favors an increase in the composition of vegetation species that are less vulnerable to the human-induced stress. This is often considered as rangeland degradation because the ecosystem configuration that emerges in response to the stress is usually economically less productive, and therefore less preferred, than the previous condition. Adam Smith's (1776) assertion that the individual "...intends only his own gain, and he is in this, as in many other cases, led by an invisible hand to promote an end which was no part of his intention" was made in the context of explaining unintended social good caused by commerce, but it also seems appropriate to apply the quote in the context of setting in motion a chain reaction of ecosystem change leading to unintended degradation.

The hydrology, vegetation and the physical and chemical characteristics of the soil aspects are inextricably linked within an ecosystem. Therefore, the associated chapters of this book which discuss aspects of vegetation degradation (Archer and Stokes, this volume), nutrient degradation (Havstad et al., this volume) and re-distribution (Ludwig and Tongway, this volume) must be considered together with this chapter on hydrologic degradation to fully understand the dynamics of ecosystem change. For example, increased grazing pressure may reduce cover, which increases the incidence of direct raindrop impact on the soil, which increases crusting, which increases runoff, which increases soil and nutrient loss. This leads to a combination of reduced vegetation production and greater concentration of grazing pressure on the remaining plants, triggering even more runoff and erosion, thus beginning another round of what can become a self-reinforcing cycle (Thurow, 1991).

The reason for stressing the human connection in a discussion of the role of hydrologic change in the degradation sequence is that natural changes in the hydrologic characteristics of a site typically occur very slowly. The soils and plants which characterize the current ecosystem would not be present if they were unable to cope with temporal variation in precipitation. Even the much-cited case of ecosystem change associated with the Sahelian droughts over the last several decades (c.f. Le Houérou, 1989) cannot be interpreted independent of the concurrent, dramatic increases in human and livestock populations which have a greatly impacted ecosystem structure and function (Cissé, 1981). In fact, when comparing the impact of droughts from the early part of the century with recent droughts, it appears that intensified land use pressure is more to blame for changing the ecosystem characteristics than the periods of low rainfall (Laya, 1975; Lamprey, 1988).

2. HYDROLOGIC ATTRIBUTES OF ECOSYSTEMS

Infiltration rate, soil water storage capacity, rain use efficiency and precipitation characteristics are hydrologic attributes vital to ecosystem structure and function (Aronson et al., 1993ab; Tongway and Hindley, this volume). Understanding how each of these attributes may be altered by human-induced activities is necessary for understanding degradation and restoration processes. Assessment of the impact on these attributes can be used as criteria of sustainability associated with a particular land use and can be used to guide management.

2.1 Infiltration rate

The rate at which water enters the soil (infiltration) is determined to a large degree by soil structure. Soil structure is determined by the size of soil particles (texture), the degree to which the soil particles are held together in individual clusters (aggregation), and the ability of those clusters to maintain their structural integrity when wetted (aggregate stability) (Boyle et al., 1989). Aggregation occurs when roots, fungal hyphae, and/or adhesive by-products of organic matter decay and microbial syntheses mechanically bind soil particles. These mechanically-bound particles are then cemented together by resistant humus components which form chemical bonds (Brady, 1974). Macropores increase as the soil becomes more aggregated (Allison, 1973; Virginia, 1986). If the soil is well aggregated and the aggregate bonds are strong, the soil will be very porous when wet, thereby resulting in a high infiltration rate. Since water availability is often a limiting factor of rangeland productivity, a high infiltration rate is desirable because water will enter the soil and be available for vegetation instead of running off site, carrying with it soil and nutrients.

Aggregate bonds are susceptible to being broken when raindrops strike the soil. Cover intercepts and dissipates the kinetic energy of raindrops, thereby protecting aggregate structure. Vegetation type affects the amount and structure of associated cover; therefore, the infiltration rate differs among vegetation types. The amount of cover, and hence the rate of infiltration, is usually greater under trees and shrubs, followed in a decreasing order by bunchgrass, shortgrass and bare ground (Hester et al., 1997). These different growth forms also influence the consistency of cover provided throughout the year, with sites dominated by woody species or bunchgrasses tending to have less variation of infiltration rates between seasons than sites dominated by herbaceous species with more seasonally variable cover characteristics (Thurow et al., 1986).

Another important difference between vegetation types relative to their effect on infiltration rate is related to the amount of litter. For example, bunchgrasses and shrubs tend to produce greater amounts of foliage than annuals and shortgrasses. The fallen foliage accumulates as litter which, in turn, leads to an increase in soil organic matter. Litter also creates a more consistent temperature and moisture microenvironment that favors microorganism activity. These factors enhance formation of stable soil aggregates.

When cover and organic matter inputs are reduced in association with a change in land use pattern, more of the soil is exposed to the kinetic energy of raindrop impact, which breaks the aggregate apart. Deterioration of aggregate stability associated with a decrease in organic matter and microorganism activity allows clay particles to disperse more easily when rapidly wetted (slaking). The combination of slaking and the breaking of aggregates associated with raindrop impact results in some of the unbound soil particles lodging in the remaining pores between aggregates, making them smaller or sealing them completely (Lynch and Bragg, 1985). This is one way in which soil crusts are formed. A "washed in" layer, where clay particles have clogged soil pores to form a crust, may reduce infiltration rate by as much as 90% (Boyle et al., 1989). Also, as vegetation cover is reduced, there is less obstruction to overland flow, thus there is greater momentum of the runoff and greater erosive energy. Machinery or livestock hoof impact may temporarily break the crust and help to incorporate mulch and seeds into the soil. However, the aid to infiltration is short-lived because the disaggregated soils will have unstable pores which quickly become plugged when wetted. The pressure associated with the impact may compress the soil and reduce pore space, thereby increasing bulk density. Also, the potential for wind erosion increases when the soil has been churned to dust. To prevent crusting it is vital that the land use system be sensitive to the need to maintain cover.

The hydrologic characteristics of various vegetation types can be expected to confer some competitive advantages to vegetation types with the greatest infiltration rates. Much of the water that flows overland does not leave the site. Rather, it flows for a short distance until it reaches an area of higher infiltration capacity that can accommodate both the falling precipitation and the overland flow. The net result is that in a vegetation mosaic the rainfall is redistributed by overland flow so that the mineral soil near some species associated with high infiltration rates receive more water than the intervening areas. In this way, areas under shrubs may capture overland flow from interspaces of grass or bare soil (Cornet et al., 1992).

Increased dominance of shrubs in a savannah where fire and grazing is controlled may sometimes result in improvement of the infiltration rate of the overall site because of the associated increase of cover and organic matter input. Humans using the land for grazing may consider the site degraded from an economic perspective, less herbaceous forage is being produced for use by livestock. Therefore, if the definition of range condition has an economic component, infiltration rate alone is not a good indicator. If, however, range condition is defined only in terms of a sustainable biomass production, infiltration rate is a good indicator because a greater infiltration rate will mean that more of a limiting factor (water) can enter the soil and become available for vegetation growth. Also, a higher infiltration rate means that there will be less loss of soil and nutrients associated with water erosion (see Archer and Stokes, this volume, for further discussion of economic vs. ecological degradation).

2.2 Soil water storage capacity

The ability to store rainwater in the friable, upper soil layers is an important determinant of plant growth patterns and overall production on a site. The amount of water able to be stored in soil is determined by the capillary attractive forces associated with the pore characteristics of the soil, the adsorption characteristics at the surfaces of the soil particles and by the depth of the soil.

The structure, organic matter content, texture, and type of clay minerals in the soil affect matrix potential, which collectively reflects the capilarity and adsorption traits of the soil. These characteristics effect drainage and the degree to which plants can extract water from the soil. The previous discussion of how vegetation types and land use influence soil structure and organic matter content, thereby influencing infiltration rate, also applies to water storage traits within the soil profile. Reduction of organic matter input and degradation of soil structure will degrade a sites' ability to retain water.

The presence of soil indicates that under past conditions the erosion rate was less than the rate of formation and/or deposition. If current land use practices disrupt the ability of the site to retain soil, the site will lose water holding capacity. The relationship between water storage capability and plant production is strongest on sites with shallow soil since these sites are especially vulnerable to the erratic rainfall which is characteristic of many arid and semi-arid rangelands (Floret and Pontanier, 1982).

The loss of soil moisture storage capability will shorten the amount of time before plants will run out of water and show signs of drought stress. As a site becomes more vulnerable to drought, the difficulty in maintaining plant cover increases and the site becomes more vulnerable to accelerated erosion, which creates a spiral of decreasing production potential (Thurow, 1991; Le Houérou, 1996). Indeed, the definition of desertification as "a diminution or destruction of biological production potential" (Dregne, 1987) is not specifically linked to precipitation (see also O. Arnalds, this volume). Therefore, susceptibility to drought may increase and a loss of production potential linked to a decrease in soil moisture storage capability may occur.

2.3 Precipitation characteristics

Amount, intensity, duration, form, and temporal and spatial distribution of precipitation are climatic variables that fundamentally influence composition and production of the vegetation community and formation of soil. Generally, climatic conditions are inherent characteristics which cannot be directly altered by land use patterns. However, in extreme cases of widespread degradation, a severe reduction of vegetation cover can change surface reflectivity, which can theoretically inhibit cloud formation and reduce precipitation (Charney et al., 1975; Bryant et al., 1990; Pielke et al., 1998). Global warming, associated with emission of greenhouse gases, is another way in which humans can indirectly alter precipitation patterns. It is hypothesized that warmer air temperatures would cause shifts in ocean currents which would alter global precipitation patterns (IPCC, 1996).

The uncertainty associated with the temporal and spatial variability of rainfall on most rangelands of the world tends to cause confusion and indecision, resulting in either inaction or *ad hoc* responses which do not fully consider the complex, long-term ecological and socio-economic interactions associated with water shortage (Wilhite and Glantz, 1985). Due to the unstable nature and complexity of atmospheric dynamics, the theoretical limit of accurate weather prediction does not exceed a few weeks (Skukla, 1985). Much attention has been devoted to search for trends or cycles of long-term climate, yet discussion of this issue in the popular press ignores the most important aspect of rainfall on many rangelands – its extreme variability. Even if trends or cycles do exist, it appears that reliable, site-specific seasonal forecasts one or more seasons in advance probably are unattainable (Rasmusson, 1987) and have limited managerial value (Glantz and Katz, 1977).

2.4 Rain use efficiency

Rain use efficiency (RUE) is a term commonly used to indicate the ratio of aboveground phytomass produced per hectare per millimeter of rainfall (Le Houérou, 1984). Factored into this term is the recognition that much of the precipitation may exit the site via runoff or deep drainage. For example, rain use efficiency would decrease as soil crusts form and runoff increases. For this reason, RUE is often used to explain production declines associated with desertification. In contrast, water use efficiency tends to be defined by agriculture and ecosystem scientists as the ratio of net primary production to evapotranspiration and by plant physiologists as the ratio of net photosynthesis to transpiration (Kramer, 1983).

The characteristics of the vegetation present on the site also determines its ability to access and use water. The photosynthetic pathway (C3, C4, CAm), the degree of stomatal control and the ability to lower xylem pressure potential to extract water from a drying soil are examples of physiological traits that influence rain use efficiency. The extent and characteristics of the root system, the amount of rainfall intercepted and evaporated back to the atmosphere without ever reaching the ground, and the amount of rainfall intercepted and concentrated via stemflow at the base of the plant are structural examples of how water access may be affected.

Plant communities with quite different rain use efficiencies can dominate a site. For example, the northwestern U.S. perennial bunchgrass community dominated by bluebunch wheatgrass (*Agropyron spicatum* (Pursh) Scribn.) cannot sustain heavy grazing pressure. In response to grazing the wheatgrass may be replaced by a dominance of either short-lived annuals such as cheatgrass *(Bromus tectorum* L.) or woody species such as sagebrush (*Artemisia tridentata* Nutt.). These three communities have quite different characteristics with regard to access and use of water, biomass production and value related to their ability to support livestock and wildlife (Blaisdell et al., 1982; Smith and Nowak, 1990).

3. MANAGEMENT CHALLENGES

Lowdermilk's (1953) classic review of civilization and natural resource use concluded that preservation of the soil resource and associated hydrologic condition is essential to the well-being of society. Nobody thinks that accelerated erosion and deterioration of hydrologic condition is good, so why is this recurrent theme of civilization decline repeated and still a problem on many rangelands of the world today?

Land use practices and the concern for soil and water conservation vary greatly depending upon whether the management time horizon is focused on short-term economic gain or long-term sustained yield. Economic logic suggests that conservation of the soil resource will occur when the discounted value of future profit potential exceeds current returns. In overpopulated regions of the world, the need for immediate survival outweighs any consideration of future productivity (see also Narjisse, this volume). Consequently, people in these overpopulated areas may indeed understand that the resource is deteriorating, but they may not be in a position to change their land use patterns since their immediate concerns are to produce enough to stay alive. In developed countries, the factors affecting conservation choices are usually not so dire; rather than literal life or death tradeoffs, land use practices which cause accelerated erosion may be prompted by the need to generate sufficiently high yields to remain financially solvent. Expedient decisions to cushion immediate financial pressures at the expense of vegetation cover ultimately lead to accelerated erosion, which leads to decreased soil moisture storage, which causes decreased production potential associated with increased frequency of drought stress. In addition to the immediate costs of loss of productivity, erosion reduces the magnitude of future benefits that could be gained through introduction of improved livestock breeds or better husbandry techniques.

It is particularly difficult to manage grazing pressure in regions characterized by high intensity storms with a variable spatial and temporal distribution. Consequently, rangelands that share these traits tend to be among the most degraded ecosystems of the world. A causal land use factor contributing to the degradation is linked to the difficulty of maintaining the herd flexibility needed to take advantage of herbaceous growth spurts while ensuring maintenance of vegetation cover required to dissipate the erosive energy of rainfall and overland flow. In such regions, communal land tenure provided pastoralists with the flexibility of movement needed to derive a sustainable livelihood from these lands. An increase in human and livestock population and/or a disruption of movement patterns by settlement and/or political boundary decisions has undermined the traditional sustainable strategy of livestock production on these lands (Narjisse, this volume). Large modern ranches can be managed to maintain the needed flexibility of grazing use, but that option is currently not politically feasible on many communal lands because implementing this form of land use would require a dramatic decrease in the people and livestock living on the range (Jahnke, 1982).

Precipitation variability is an important characteristic of many rangeland climates. It is disingenuous to use the unpredictability of precipitation as an excuse for inadequate planning decisions that protect the vegetation and soil, and thereby maintain hydrologic condition. Exposing the land to accelerated erosion hazard should be viewed as a managerial failure, instead of making precipitation variability a scapegoat for faulty policies or management (see also Narjisse, this volume). One of the reasons that policy-makers and landowners persist in treating drought as a quirk of nature is that if they accept the challenge of planning for drought, then they implicitly accept the risk and responsibility for the consequences of a proactive set of responses to drought. These are difficult responsibilities for policy-makers and pastoralists to bear because the costs of planning for drought are fixed and occur now (i.e., implementation of a rapid de-stocking policy or the maintenance of reserve pasture) while the costs of degradation from drought are uncertain and will occur later. Furthermore, costs such as productivity losses due to soil erosion are difficult or impossible to incorporate into short-term analyses because these costs are primarily manifest over the long-term.

In sum, the challenges to maintaining good hydrologic condition has intertwined economic and ecological roots. The ecological signals regarding long-term sustainability and the cumulative economic consequences are often not sufficiently clear, early on, to promptly influence pastoral decisions about land use. Government representatives, who should be taking a long-term societal view of sustainable resource management, are reluctant to place erosion control as top priority of land management because enforcement of such an objective "is thorny, it is packed with political dynamite, and it will always keep for another couple of years" (Huxley, 1937). Management and policy tools must improve the integration of economic and ecological aspects of natural resource management, especially by including the long-term, irreversible costs of erosion as part of the financial cost-benefit analysis of an activity (Walker, 1993). Defining the management and policy analysis unit at watershed scale of resolution facilitates a more complete accounting of the serial downstream costs and benefits associated with land use activities than smaller planning units. A watershed analysis units allows the upstream and downstream activities to be linked, therefore providing a rationale that makes investments in management and restoration efforts of uplands more effective and economically attractive to downstream interests (see Ludwig and Tongway, this volume, for 'landscape' perspective).

4. RESTORATION

Severely degraded rangelands are not easily improved on sites where the water and nutrient cycles have been disrupted (Westoby et al., 1989; Milton et al., 1994). The self-generating nature of degradation processes often means that restoration cannot be achieved by simply removing the stress that started the degradation process (Friedel, 1991; Laycock, 1991; Daily, 1995; Rietkerk et al., 1996; Archer and Stokes, this volume). Consequently, degraded landscapes are often abandoned (Barrow, 1991) because restoration prescriptions often require management expenditures that are unlikely to be recouped in terms of on-site production improvement.

A consideration that is central to any restoration effort is the definition of management scale. A watershed is a logical, natural management unit for restoration activities (Thurow and Juo, 1995; see Ludwig and Tongway, this volume, for 'landscape' perspective). The drainage patterns of a watershed form the framework of important energy flow and nutrient cycles that determine the structure and function of the ecosystem. These flow patterns are also an important context in which investment in restoration efforts should be evaluated. The value of restoring a degraded area should not only be measured in terms of improved on-site production (as is usually the case), but also in terms of the future costs to both on-site and downstream productivity if the site is allowed to continue to degrade. A watershed provides a management unit that is large enough to internalize restoration benefits that extend downstream.

Even if the opportunity costs and shadow costs to the entire watershed are considered, the scale of degradation and the economic realities of arid and semiarid ecosystems prevent widespread adoption of restoration strategies requiring intensive management. Since funding for restoration is generally very difficult to procure and sustain, it is desirable to devise focused interventions that can jump-start a repair process which will thereafter be self-generating (autogenic) (Whisenant et al., 1995).

Plant-soil interactions serve as one of the most influential positive feedback loops in rangeland systems (Graetz, 1991). The rate of recovery of damaged ecosystems is often governed by the degree to which biotic components can ameliorate the soil and the associated water and nutrient flow patterns (Tongway and Hindley, this volume). In the long-term, hydrologic processes controlled by soil surface conditions are only repaired and maintained by increasing plant production and protecting the soil surface with plant litter or living vegetation. Often, the greatest challenge to initiating positive feedback loops of the restoration process is the difficulty of re-establishing conditions which are needed to foster vegetation survival and growth.

On severely degraded sites it may be necessary to concentrate water and nutrients so that vegetation has a chance to thrive and thereby start the soil repairing process (Ludwig and Tongway, this volume). One way to accomplish this is by altering microtopographic features, such as installing microcatchment basins, that facilitate capture and accumulation of water, soil, nutrients, organic matter and propagules (Reij et al., 1988). In areas with high wind erosion, simply placing cut branches on crusted soil may enable significant eolian dust deposition (Chase and Boudouresque, 1987) which has high water holding capacity and is rich in organic matter and nutrients (Drees et al., 1993). Planting keystone species with desirable characteristics, such as rapid growth and nitrogen fixation capability, into the area of concentrated water and nutrients can accelerate further amelioration of microenvironmental conditions and speed plant succession (Manu et al., 1994; Whisenant et al., 1995).

Initiating dynamic vegetation change in patches stimulates biological, edaphic, hydrologic and microenvironmental processes which gradually benefit adjacent parts of the landscape (Ludwig and Tongway, this volume). This sets in motion a range of serial ecological and economic benefits to downstream portions of the watershed which are aided by the upland restoration efforts. In this way, a restoration effort that targets a few destabilization points may have far reaching effects. For example, striped vegetation patterns composed of densely vegetated patches regularly alternating with almost bare areas commonly occur on gentle smooth slopes of semi-arid lands (Belsky, 1986; Montana, 1992; Ludwig and Tongway, 1995). The existence of these patterns can generally be explained by the positive feedback of plant density and water infiltration (Belsky, 1986) and/or nutrient retention (Ludwig and Tongway, 1995). In south-central Niger, when heavy fuel wood harvest and grazing pressure caused gaps to form in the vegetation stripes, runoff and sediment flowed unchecked through the degraded thicket gaps. This resulted in widespread degradation of the laterite plateau on which the stripped mosaic existed and on downslope portions of the landscape (Manu et al., 1994).

Restoration of the landscape was initiated by targeting efforts on the relatively small portion of the landscape represented by the degraded thicket gaps. Survival of tree seedlings planted in the crusted soils in these gaps was 0%. However, survival of seedlings was 99% when microcatchments were installed to aid concentration of water and nutrient necessary for plant establishment. Two years after being planted, hardy keystone species such as *Acacia holosericea* A. Cunn. ex. G. Don had grown to a height of 2.2 m. The deposition of litter and alteration of microclimate caused by these keystone species created a microenvironment that fostered microorganism activity and natural plant succession. These changes also resulted in an amelioration of the soil that significantly increased the infiltration rate. This led to a cessation of water flowing off the plateau, removing the source of erosion hazard both on the plateau and in downstream areas. This practice has being voluntarily initiated by local farmers in the region as a cost-effective means for reclaiming degraded landscapes that threaten their livelihood (Manu et al., 1994).

5. SUMMARY

Infiltration rate, water storage soil capability, precipitation patterns and rain-use efficiency are important hydrologic attributes which influence the nature of ecosystems. A change in ecosystem structure and function will occur when human land use patterns disrupt an existing stable linkage of water/soil/vegetation. A disruptive change may or may not result in less biomass being produced, but it is likely to be considered degradation because the site will be less able to support the plant species intolerant of the land use pressure that caused the change. Restoration of a denuded environment may require implementation of techniques designed to concentrate water and nutrient resources so that vegetation can thrive, thereby initiating a positive feedback loop. Targeted applications of these techniques aimed at altering hydrologic and nutrient dynamics can help to restore stability on the landscape.

REFERENCES

Allison, F.E. 1973. Soil Organic Matter and its Role in Crop Production. Elsevier, New York, NY.

Aronson, J., Floret, C., Le Floc'h, E., Ovalle, C. and Pontanier, R. 1993a. Restoration and rehabilitation of degraded ecosystems in arid and semi-arid lands. I. A view from the south. Restoration Ecology 1, 8–17.

Aronson, J., Floret, C., Le Floc'h, E., Ovalle, C. and Pontanier, R. 1993b. Restoration and rehabilitation of degraded ecosystems in arid and semi-arid lands. II. Case studies in southern Tunisia, central Chile and northern Cameroon. Restoration Ecology 1, 168–187.

Barrow, C.J. 1991. Land Degradation. Cambridge University Press, New York, NY.

Belsky, A.J. 1986. Population and community processes in a mosaic grassland in the Serengeti, Tanzania J. Ecol. 74, 841–856.

Blaisdell, J.P., Murray, R.B. and McArthur, E.D. 1982. Managing intermountain rangelands: sagebrush-grass ranges. General Technical Report INT-134. USDA Intermountain Forest and Range Experiment Station, Ogden, Utah.

Boyle, M., Frankenberger, Jr., W.T. and Stolzy, L.H. 1989. The influence of organic matter on soil aggregation and water infiltration. J. Prod. Agr. 2, 290–299.

Brady, N.C. 1974. The Nature and Properties of Soils. 8th edn. Macmillan Publishers, New York, NY.

Bryant, N.A., Johnson, L.F., Brasel, A.J., Balling, R.C., Hutchinson, C.F. and Beck, R. 1990. Measuring the effect of overgrazing in the Sonoran Desert. Climatic Change 17, 243–264.

Charney, J., Stone, P.H. and Quirk, W.J. 1975. Drought in the Sahara: a biophysical feedback mechanism. Science 187, 434–435.

Chase, R.G. and Boudouresque, E. 1987. Methods to stimulate plant regrowth on bare Sahelian forest soils in the region of Niamey, Niger. Agriculture, Ecosystems and the Environment 18, 211–221.

Cissé, S. 1981. Sedentarizations of nomadic pastoralists and pastoralization of cultivators in Mali. In: The Future of Pastoral People. Ed. D. Aronson. CDRI, Ottawa, Ontario.

Cornet, A.F., Montana, C., Delhoume, J.P. and Lopez-Portillo, J. 1992. Water flows and the dynamics of desert vegetation stripes. In: Landscape Boundaries: Consequences for Biotic Diversity and Ecological Flows. Eds. A.J. Hansen and F. di Castri. pp. 327–345. Springer-Verlag, New York, NY.

Daily, G.C. 1995. Restoring value to the world's degraded lands. Science 269, 350–354.

Dregne, H.E. 1987. Reflections on the U.N. plan of action to combat desertification. Desertification Control Bull. 15, 8–11.

Drees, L.R., Manu, A. and Wilding, L.P. 1993. Characterization of aeolian dusts in Niger, West Africa. Geoderma 59, 213–233.

Floret, C. and Pontanier, R. 1982. L'aridité en Tunisie présaharienne. Climat, sol, végétation et aménagement. Travaux et Doctorat de l'ORSTOM, no. 150.

Friedel, M.H. 1991. Range condition assessment and the concept of thresholds: A viewpoint. J. Range Management 44, 422–426.

Glantz, M.H. and Katz, R.W. 1977. When is a drought a drought? Nature 267, 192–193.

Graetz, R.D. 1991. Desertification: a tale of two feedbacks. In: Ecosystem Experiments. Eds. H.A. Mooney, E. Medina, D.W. Schindler, E.D. Schulze and B.H. Walker. pp. 59–87. Wiley, Chichester.

Hester, J.W., Thurow, T.L. and Taylor, Jr., C.A. 1997. Hydrologic characteristics of vegetation types as affected by prescribed burning. J. Range Management 50, 199–204.

Holling, C.S. 1973. Resilience and stability of ecological systems. Annual Review of Ecology and Systematics 4, 1–23.

Huxley, E. 1937. The menace of soil erosion. J. Royal African Soc. 36, 365–370.

Intergovernmental Panel on Climate Change 1996. Climate Change 1995 – Impacts, Adaptations and Mitigation of Climate Change: Scientific-Technical Analysis. Cambridge University Press, New York, NY.

Jahnke, H.E. 1982. Livestock Production Systems and Livestock Development in Tropical Africa. Kieler Wissenschaftsverlag Vauk, Kiel.

Kramer, P.J. 1983. Water Relations of Plants. Academic Press, New York, NY.

Lamprey, H.F. 1988. Report on the desert encroachment reconnaissance in northern Sudan. Desertification Control Bulletin 17, 1–7.

Laya, D. 1975. A l'écoute des paysans et des éleveurs au Sahel. Environnement Africain 1, 53–101.

Laycock, W.A. 1991. Stable states and thresholds of range condition on North American rangelands: a viewpoint. J. Range Management 44, 427–433.

Le Houérou, H.N. 1984. Rain use efficiency: A unifying concept in arid-land ecology. J. Arid Environ. 7, 1–35.

Le Houérou, H.N. 1989. The Grazing Land Ecosystem of the African Sahel. Springer Verlag, Berlin.

Le Houérou, H.N. 1996. Climate change, drought and desertification. J. Arid Environ. 34, 133–185.

Lowdermilk, W.C. 1953. Conquest of the land through 7,000 years. U.S. Dep. of Agr. – Soil Conserv. Ser., Agr. Info. Bull. No. 99.

Ludwig, J.A. and Tongway, D.J. 1995. Spatial organization of landscapes and its function in semi-arid woodlands, Australia. Landscape Ecology 10, 51–63.

Lynch, J.M. and Bragg, E. 1985. Microorganisms and soil aggregate stability. Adv. Soil Sci. 2, 133–171.

Manu, A., Thurow, T.L., Juo, A.S.R., Zanguina, I., Gandah, M. and Mahamane, I. 1994. Sustainable land management in the Sahel: A case study of an agricultural watershed at Hamdallaye, Niger. TropSoils CRSP Bulletin No. 94–01. Texas A&M University, College Station, Texas.

May, R.M. 1977. Thresholds and breaking points in ecosystems with a multiplicity of steady states. Nature 269, 471–477.

Milton, D.J., Dean, W.R.J., du Plessis, M.A. and Siegfried, W.R. 1994. A conceptual model of arid rangeland degradation. BioScience 44, 70–76.

Montana, C. 1992. The colonization of bare areas in two-phase mosaics of an arid ecosystem. J. Ecol. 80, 315–327.
Pielke, R.A., Avissar, R., Raupach, M., Dolman, A.J., Zeng, X.B. and Denning, A.S. 1998. Interactions between the atmosphere and terrestrial ecosystems: influence on weather and climate. Global Change Biology 4(5), 461–475.
Rasmusson, E.M. 1987. Global prospects for the prediction of drought: A meteorological perspective. In: Planning for Drought: Toward a Reduction of Societal Vulnerability. Eds. D.A. Wilhite and W.E. Easterling. pp. 31–44. Westview Press, Inc., Boulder, CO.
Reij, C., Mulder, P. and Begemann, L. 1988. Water harvesting for plant production. World Bank Technical Paper No. 91. The World Bank, Washington, DC.
Rietkerk, M., Ketner, P., Stroosnijder, L. and Prins, H.H.T. 1996. Sahelian rangeland development: a catastrophe? J. Range Management 49, 512–519.
Rietkerk, M. and Vandekoppel, J. 1997. Alternate stable states and threshold effects in semi-arid grazing systems. Oikos 79, 69–76.
Skukla, J. 1985. Predictability. In: Issues in Atmospheric and Oceanic Modeling Part B: Weather Dynamics. Ed. S. Manabe. pp. 87–122. Academic Press, New York, NY.
Smith, A. 1776. An Inquiry into the Nature and Causes of the Wealth of Nations. Book IV, Chapter ii. W. Strahan & T. Cadell Publishers, London, p. 456.
Smith, R.L. 1974. Ecology and Field Biology. Harper & Row, New York, NY.
Smith, S.D. and Nowak, R.S. 1990. Physiological ecology of plants in the intermountain lowlands. In: Ecological Studies, Vol. 80. Plant Biology of the Basin and Range. Eds. C.B. Osmond, L.F. Pitelka and G. Hidy. pp. 179–243. Springer-Verlag, Berlin.
Thurow, T.L., Blackburn, W.H. and Taylor, Jr., C.A. 1986. Hydrologic characteristics of vegetation types as affected by livestock grazing systems, Edwards Plateau, Texas. J. Range Management 39, 505–509.
Thurow, T.L. 1991. Hydrology and erosion. In: Grazing Management: An Ecological Perspective. Eds. R.K. Heitschmidt and J.W. Stuth. pp. 141–159. Timber Press. Portland, OR.
Thurow, T.L. and Juo, A.S.R. 1995. The rationale for using a watershed as the basis for planning and development. In: Agriculture and Environment: Bridging Food Production and Environmental Protection in Developing Countries. Eds. A.S.R. Juo and R.D. Freed. pp. 93–116. American Society for Agronomy Special Publication No. 60. Madison, WI.
Virginia, R.A. 1986. Soil development under legume tree canopies. Forest Ecology and Management 16, 69–79.
Walker, B.H., 1993. Stability in rangelands: ecology and economics. In: XVII International Grassland Congress, Vol. III. Eds. M.J. Baker, J.R. Crush and L.R. Humphreys. pp. 1885–1890. Keeling & Monday Ltd., New Zealand / Australia.
Westoby, M., Walker, B. and Noy-Meir, I. 1989. Opportunistic management for rangelands not at equilibrium. J. Range Management 42, 266–274.
Whisenant, S.G., Thurow, T.L. and Maranz, S.J. 1995. Initiating autogenic restoration on shallow semiarid sites. Restoration Ecology 3, 61–67.
Wilhite, D.A. and Glantz, M.H. 1985. Understanding the drought phenomenon: the role of definitions. Water Int. 10, 111–120.

Erosion models: use and misuse on rangelands

Frederick B. Pierson, Jr.
United States Department of Agriculture, Agricultural Research Service, 800 Park Blvd, Plaza 4, Suite 105, Boise, Idaho, USA 83712
Tel: 208 422 0720; Fax: 208 334 1502; E-mail: <fpierson@nwrc.ars.pn.usbr.gov>

ABSTRACT Erosion modeling is currently of limited use in combating rangeland desertification. Empirical (e.g., RUSLE) and mechanistic (e.g., WEPP) erosion models exist, but are primarily designed for cropland systems and are difficult to adapt for use on rangelands. This is due to a critical lack of erosion data representing rangeland ecosystems. In particular, the feedback between changes in vegetation characteristics, hydrology and erosion processes is not well understood. For rangeland erosion modeling to move forward, improved databases and model parameter estimation techniques that represent vegetation, soil, and management-induced spatial and temporal variability are needed for model building, testing and implementation. To better address global desertification issues, erosion modeling must move beyond general or simplistic long-term erosion estimates. Future erosion models must be usable in an ecological context and address "state and transition" and "threshold" concepts and principles.

Key words: desertification, erosion, modeling, rangeland, RUSLE, WEPP.

1. INTRODUCTION

Rangelands are a vast, complex, and diverse resource, including many ecosystems that are fragile and sensitive to disturbance. Dry climates combined with highly variable soil conditions produce extreme spatial and temporal variation in rangeland vegetation that is very slow to respond to management (Branson et al., 1981). A great number of rangelands throughout the world have experienced a loss of site productivity through changes in vegetation composition, productivity and structure, resulting in increased erosion loss of productive top soil. This process, known as rangeland desertification, is a complex ecological problem controlled by abiotic and biotic factors and driven by land use impacts, drought, fire and other disturbances (Dodd, 1994; Mainguet, 1991).

The complexity of desertification problems has compelled rangeland managers to seek tools to assist them in making management decisions. In the absence of tools that address the ecological processes in desertification, managers have adapted existing tools designed to predict long-term erosion rates for crop production systems. The application of cropland-based models to rangelands, with little change to accommodate the unique ecological aspects of rangeland systems, has led to erroneous management and policy decisions (Pierson et al., 1996). The experience has left natural resource managers wary of modeling as a practical problem-solving tool (Wagenet, 1988).

2. EROSION PREDICTION ON RANGELANDS

In the 1960s, mathematical computer simulation modeling efforts in the United States, driven by soil conservation programs, began to focus on erosion prediction under cropland agriculture (Thurow, 1996). The concept was to predict long-term annual erosion rates and compare them to the inherent capability of a soil to regenerate. Soil loss tolerance values (t) could then be computed and conservation practices employed to limit erosion to less than the estimated t-value.

Empirical modeling has dominated erosion prediction technology because of its simplicity and ease of use. A number of empirical approaches have been used to predict erosion, but the most widely accepted has been the Universal Soil Loss Equation (USLE) (Wischmeier, 1976). The USLE was based on over 10,000 plot years of cropland data and was designed to estimate sheet and rill erosion throughout a watershed (Meyer, 1981). USLE technology made no specific provisions for rangeland applications until the recent Revised Universal Soil Loss Equation (RUSLE) became available. RUSLE is an improvement over USLE for use on rangeland, but is still restricted to the simple linear empirical form of the USLE by using the same by-product of six factors (Lane et al., 1992; Renard et al., 1997; Weltz et al., 1996).

In RUSLE, each of the six factors is calculated as a function of subfactors reflecting current knowledge of soil erosion science (Lane et al., 1992). The cover and management factor (C) is the critical term in RUSLE for rangelands, since it is directly affected by land management practices. The C-factor varies from near 0, for a dense grass area with no exposed bare soil, to 1.5 for freshly tilled and bedded soil. For rangelands, the C-factor varies as a function of prior land use, canopy cover, soil cover and soil surface roughness, and does not change over the simulation period. The prior land use sub-factor is based on the time since last disturbance, root biomass, and organic material in the upper 100 mm of the soil. The canopy cover subfactor is related to the fractional cover of the soil surface provided by aboveground plant biomass and the height that raindrops fall after leaving a plant and impacting the soil surface. The soil surface cover sub-factor estimates the portion of the soil surface covered by non-eroding material (basal area of plants, rocks, and organic litter). The surface roughness factor is based on the random roughness of the soil surface and the root biomass in the upper 100 mm of the soil (Weltz et al., 1996).

RUSLE is a conceptual improvement over USLE for use on rangelands in that it incorporates a number of additional factors that influence erosion on rangelands. However, the data used to develop such algorithms are limited and there has been little to no validation (Lane et al., 1992). To be useful as a rangeland management tool, the empirical relationships in RUSLE need to statistically represent a wider array of rangeland soil-vegetation-management scenarios. The model also needs rigorous testing using independent data sets to determine where and when the model is truly applicable to resource management problems.

Recent attempts to explicitly model soil erosion have used a mechanistic approach. An example is the Water Erosion Prediction Project (WEPP) model developed by the United States Department of Agriculture, Agricultural Research Service (USDA-ARS) (Flanagan and Nearing, 1995; Laflen et al., 1991). The WEPP model is intended to apply to all rangeland and pastureland situations where water erosion occurs, including that resulting from rainfall, snowmelt, irrigation, and ephemeral gullies (Weltz et al., 1996). The objective of WEPP was to replace the USLE and RUSLE erosion models with a "new generation" water erosion prediction technology (Laflen et al., 1991; Lane et al., 1992).

WEPP is a process-based erosion model that operates on a daily time step and incorporates temporal changes in soil erodibility, management practices, above- and belowground biomass, litter biomass, plant height, and canopy and ground cover in the prediction of soil erosion. Linear and nonlinear slope segments and multiple soil series and plant communities can be represented with the model. The WEPP model operates at two scales: hillslope and watershed. The hillslope model has seven conceptual components: climate, topography, soils, hydrology, erosion, management, and plant growth and decomposition. The two components that rangeland managers can most readily manipulate are the plant and management components. Plant growth is simulated as a function of temperature, solar radiation, and soil water content. The growth rate of aboveground biomass for rangeplant communities is simulated using potential-growth curve, defined with either a unimodal or a bimodal distribution of plant growth (Arnold et al., 1995). Plant parameters calculated by daily simulation include canopy height and cover, aboveground standing biomass, plant density, leaf area index, litter mass and cover, basal plant cover, rock and cryptogam cover, total ground cover, root biomass, and root distribution with depth (Weltz et al., 1998).

The WEPP model provides a choice of four rangeland management options: grazing, fire, herbicide application, and complete protection. The user can define the type, severity, and timing of the management activity to be simulated. The effect of grazing is represented by removal of standing biomass with a corresponding reduction of canopy and basal plant cover and the transfer of standing dead biomass to litter. Fire and herbicide applications are represented by the amount and timing of removal of biomass from some portion of the hillslope. A hillslope within the WEPP model can have as many as ten combinations of soil types, vegetation communities, or management activities (Weltz et al., 1998).

At the watershed scale, the WEPP model links multiple hillslopes together. Such versatility allows the user to represent a wide range of management practices and estimate soil loss and deposition from any location within a watershed. Natural resource planners can evaluate resource management systems based on: (1) overall soil loss; (2) location of soil loss and deposition; and (3) off-site sediment yield that may result in decreased water quality. The watershed model can be used in project planning to estimate or evaluate such things as: (1) watershed erosion control measures; (2) rate of catchment sedimentation; (3) impact of grazing systems on soil loss; and (4) soil loss from mine reclamation activities (Ascough et al., 1997).

Mechanistic models, such as WEPP, allow greater flexibility in representation of rangeland conditions and scenarios than do empirical models such as RUSLE or USLE. The output provides detailed spatial and temporal information useful to rangeland managers, but at a cost of requiring a great deal more information to parameterize and run. The WEPP technology is very young and has yet to be fully tested.

3. CHALLENGES TO EROSION MODELING ON RANGELANDS

To address desertification issues, rangeland managers need much more than the ability to predict long-term erosion rates compared to soil t-values. They need the ability to solve a wide array of resource problems and predict the possible ecological, hydrological and erosion consequences of different management alternatives (Carlson et al., 1992; Wight, 1988). They need the ability to make both short- and long-term predictions, evaluate and choose the best management alternative and assess the risk associated with making a particular management decision (Wight and Hanks, 1981). This presents a number of challenges for experimentalists and model developers alike. Basic process-level research is needed and extensive databases and new modeling approaches need to be developed.

3.1 Processes and feedback mechanisms

Future erosion prediction technology must be capable of simulating the complex interactions between vegetation characteristics, surface soil properties and hydrologic and erosion processes on rangelands (Blackburn et al., 1992; Blackburn and Pierson, 1994). In addition to vegetation biomass, changes in the structure, life-form and distribution pattern of the vegetation across the landscape, and the resultant feedback on infiltration capacity, runoff patterns and soil erodibility must be addressed (Ludwig and Tongway, 1995; Pierson et al., 1994; Spaeth et al., 1996; Ludwig and Tongway, this volume; Thurow, this volume). The association of microbiotic and physical soil crusts with vegetation and their sensitivity to management impacts need to be represented for accurate predictions of infiltration and erosion dynamics (Dobrowolski, 1994). Better representation of the temporal dynamics of soil erodibility related to soil freezing, wetting and drying and organic matter decomposition are also needed (Seyfried and Flerchinger, 1994).

3.2 Data bases

At present, erosion modeling on rangelands is data limited. A commitment is needed to collect greater amounts of high-quality data from a variety of rangeland systems. Data is vital to each step of the process of problem definition, model building, model testing and technology implementation (Pierson et al., 1996). Data is needed to quantify factors controlling the spatial and temporal variability in erosion processes at varying scales within different rangeland systems (e.g., Imeson and Cammeraat, this volume; Ludwig and Tongway, this volume). In particular, the interactions of vegetation and surface soil properties and their effect on the erosion process must be better understood and addressed in a modeling context (Blackburn et al., 1992; Blackburn and Pierson, 1994).

Effective implementation of erosion models for use in rangeland management will require relational databases that contain climate, soils, topography, land-use, management-practice, and vegetation data. In particular, associated biological inputs including plant height and density, canopy diameter, aboveground standing and litter biomass, rock cover, cryptogam cover, random roughness, and root biomass will be needed (Weltz et al., 1998). Such databases must be developed using consistent procedures and formats to allow models to be applied uniformly across all user groups and applicable geographic scales and to avoid duplication of time and effort. Technology for predicting erosion is improving, but advances are being constrained by the inability of users to estimate model parameters. Improved databases and improvements in model parameter estimation techniques which represent vegetation, soil, and management-practice induced spatial and temporal variability are needed before the full potential of erosion modeling capabilities can be achieved (Blackburn et al., 1992).

3.3 Model testing and use

Model testing is necessary to ensure that a model operates as expected, to point out problems and needed enhancements, and to demonstrate the applicability and reliability of model results for management purposes. Proper model testing involves assessing model performance within varying rangeland systems, where each test requires not only data to parameterize and run the model, but data to check predictions of all processing and output variables as well (Thurow, 1996). This involves collecting appropriate data over the entire space and time domain of each model simulation. Today's researchers often have neither the time nor resources to carry out such field data collection efforts.

The reward for testing models is generally low and the glamour and reward structure of modeling is geared to model building, not to the time-consuming and difficult tasks of model testing, sensitivity analysis, and model refinement (Thurow, 1996). The general result is to shortcut the process by sacrificing the costly and tedious testing and sensitivity analysis phase of model development. This leads to an incomplete understanding of the science and a misguided impression of the value of scientific intuition, resulting in misapplication of modeling technology (Wagenet, 1988). Researchers and model users must work cooperatively on collecting needed data for model development, systematic testing, refinement, and widespread application.

3.4 Thresholds and risk assessment

The National Research Council (1994) defined the concept of "rangeland health" as "...the degree to which the integrity of the soil and ecological processes of rangeland ecosystems are sustained." The definition is based on the concepts of "stable states" and "thresholds" derived by the balance between vegetation, soil properties and climate driven by natural events (e.g., weather, fire) or by management actions (e.g., grazing, destruction or introduction of plants) or a combination of the two (Friedel, 1991; Laycock, 1991; Archer and Stokes, this volume). These concepts are central to the problem of rangeland desertification throughout the world. Loss of topsoil and the subsequent loss in site productivity can define a threshold between two vegetation states (e.g., Figs. 3 and 4 in Tongway and Hindley, this volume), and has been cited as one of the most important factors in the desertification process (Davenport et al., 1998; Mainguet, 1991; National Research

Council, 1994; Thurow, this volume). The future challenge for rangeland erosion modeling is to aid in the process of defining thresholds and assessing the risk of crossing a threshold between different ecological states. This requires more than a long-term erosion estimate compared to a soil t-value. It requires the ability to identify an ecosystem's vulnerability to extreme climatic events (e.g., drought, high-intensity rainfall) before severe resource damage has occurred. As changes in vegetation are realized, tools are needed which can simulate the impact of those vegetation changes on runoff and erosion processes.

4. CONCLUSIONS

To date, erosion modeling has been of limited benefit to rangeland managers in their efforts to combat rangeland desertification. Existing erosion models generally do not provide what managers need to address rangeland resource problems. A comprehensive ecologically based tool that directly addresses field-scale problems on rangelands does not exist. Current models are primarily research oriented and designed to represent cropland systems and do not adequately address important rangeland processes.

Empirical erosion models are attractive due to their simplicity and cost effectiveness, but have been widely misused on rangelands. Their robustness is limited by the scope of variation in erosion response present in the data upon which the models are built. Such models may perform poorly under conditions not represented in the development database. Empirical models also do a poor job of predicting extreme system responses as extreme events are generally not well represented (Ferreira and Smith, 1988). The need for accurate predictions under extreme events is particularly important within a management or ecological context in terms of recharge, overland flow, sediment loss, etc. However, the simplicity and ease of use of empirical models will continue to entice managers and policy makers to attempt to extrapolate their use to rangelands.

Mechanistic models promise more robust erosion predictions. Future challenges for mechanistic modeling involve representing spatial and temporal variability in physical processes at different scales. Dealing with these temporal and spatial issues within the context of field research and/or model development is complex and poses a great intellectual challenge to watershed scientists. To be successful, mechanistic models must represent the vegetation-induced spatial patterns in surface soil properties across a landscape in order to adequately predict infiltration, surface runoff, and soil erosion as related to management of the landscape.

The applicability of erosion prediction technology to rangeland management would be enhanced by a change in the technology development process. Most erosion prediction tools originated as research models and were never designed to solve specific land management problems. Better management tools can be developed if model developers and users work together on an *a priori* definition of the problem to be solved. Only then can an appropriate model be developed to meet the users' needs. Future technology development efforts need to focus more on developing solutions to problems and less on adapting research-based models to resource issues.

Erosion modeling must begin to address new "state and transition" approaches to assessing and managing rangeland systems. Current resource assessment tools being developed only examine attributes of the current state of an ecosystem. Land managers also need tools that address what will happen to that same system under future management regimes and climatic conditions. They need the ability to quantify the short- and long-term impact of vegetation changes and management on soil and water resources. Help is needed in defining thresholds of erosion that, when crossed, will permanently degrade site productivity as well as in determining what management practices can be employed to protect against further degradation or to improve site productivity.

REFERENCES

Arnold, J.G., Weltz, M.A., Alberts, E.E. and Flanagan, D.C. 1995. Plant growth component. Chapter 8. In: USDA-Water Erosion Prediction Project: Technical Documentation. Eds. D.C. Flanagan and M.A. Nearing. pp. 8.1–8.41. NSERL Report. 10. National Soil Erosion Research Laboratory, USDA, Agricultural Research Service, West Lafayette, IN.

Ascough, J.C., Baffaut, C., Nearing, M.A. and Liu, B.Y. 1997. The WEPP watershed model: I. hydrology and erosion. Trans. Amer. Soc. Agric. Eng. 40, 921–933.

Blackburn, W.H. and Pierson, F.B. 1994. Sources of variation in erosion on rangelands. In: Variability of Rangeland Water Erosion Processes. Eds. W.H. Blackburn, F.B. Pierson, G.E. Schuman and R.E. Zartman. pp. 1–9. Soil Sci. Soc. Am. Special Publication 38. Soil Science Society of America, Madison, WI.

Blackburn, W.H., Pierson, F.B., Hanson, C.L., Thurow, T.L. and Hanson, A.L. 1992. The spatial and temporal influence of vegetation on surface soil factors in semiarid rangelands. Trans. Amer. Soc. Agric. Eng. 35, 479–486.

Branson, F.A., Gifford, G.F., Renard, K.G. and Hadley, R.F. 1981. Rangeland Hydrology. Range Science, Series 1, Society for Range Management, Denver, CO, 340 pp.

Carlson, D.H., Thurow, T.L. and Jones, C.A. 1992. Biophysical simulation models as a foundation of decision support systems. In: Decision Support Systems for Grazed Ecosystems: Emerging Issues. Ed. J.W. Stuth. pp. 36–66. UNESCO-MAB Book Series, Vol. 11, Parthenon Press, New York, NY.

Davenport, D.W., Breshears, D.D., Wilcox, B.P. and Allen, C.D. 1998. Viewpoint: Sustainability of pinon-juniper ecosystems – a unifying perspective of soil erosion thresholds. J. Range Manage. 51, 231–240.

Dobrowolski, J.P. 1994. In situ estimation of effective hydraulic conductivity to improve erosion modeling for rangeland conditions. In: Variability of Rangeland Water Erosion Processes. Eds. W.H. Blackburn, F.B. Pierson, G.E. Schuman and R.E. Zartman. pp. 83–92. Soil Sci. Soc. Am. Special Publication 38. Soil Science Society of America, Madison, WI.

Dodd, J.L. 1994. Desertification and degradation in sub-Saharan Africa: The role of livestock. BioScience 44, 28–34.

Ferreira, V.A. and Smith, R.E. 1988. The limited physical basis of physically based hydrologic models. In: Modeling Agricultural, Forest, and Rangeland Hydrology, pp. 10–18. ASAE Pub. 07-88, American Society of Agricultural Engineers, St. Joseph, MI.

Flanagan, D.C. and Nearing, M.A. (Eds.) 1995. USDA-Water Erosion Prediction Project: Technical Documentation. NSERL Report. 10. National Soil Erosion Research Laboratory, USDA, Agricultural Research Service, West Lafayette, IN, 276 pp.

Friedel, M.H. 1991. Range condition assessment and the concept of thresholds: a viewpoint. J. Range Manage. 44, 422–426.

Laflen, J.M., Lane, L.J. and Foster, G.R. 1991. WEPP: A new generation of erosion prediction technology. J. Soil Water Cons. 46, 30–34.

Lane, L.J., Renard, K.G., Foster, G.R. and Laflen, J.M. 1992. Development and application of modern erosion prediction technology – The USDA experience. Aust. J. Soil Res. 30, 893–912.

Laycock, W.A. 1991. Stable states and thresholds of range condition on North American rangelands: a viewpoint. J. Range Manage. 44, 427–433.

Ludwig, J.A. and Tongway, D.J. 1995. Spatial-organization of landscapes and its function in semiarid woodlands, Australia. Landscape Ecol. 10, 51–63.

Mainguet, M. 1991. Desertification: Natural Background and Human Mismanagement. Springer-Verlag, New York, NY.

Meyer, L.D. 1981. Soil erosion research leading to development of the Universal Soil Loss Equation. In: Agriculture Reviews and Manuals, ARW-W-26. pp. 1–16. USDA, Agricultural Research Service, Washington, DC.

National Research Council 1994. Rangeland Health: New Methods to Classify, Inventory, and Monitor Rangelands. Committee on Rangeland Classification, Board of Agriculture. National Academy Press, Washington DC.

Pierson, F.B., Blackburn, W.H., Van Vactor, S.S. and Wood, J.C. 1994. Incorporating small scale spatial variability into predictions of hydrologic response on sagebrush rangelands. In: Variability of Rangeland Water Erosion Processes. Eds. W.H. Blackburn, F.B. Pierson, G.E. Schuman and R.E. Zartman. pp. 23–34. Soil Sci. Soc. Am. Special Publication 38. Soil Science Society of America, Madison, WI.

Pierson, F.B., Spaeth, K.E. and Weltz, M.A. 1996. The use of models as rangeland management decision aids. In: Grazingland Hydrology Issues: Perspectives for the 21st Century. Eds. K.E. Spaeth, F.B. Pierson, M.A. Weltz and G. Hendricks. pp. 117–124. Society for Range Management, Denver, CO.

Renard, K.G., Foster, G.R., Weesies, G.A., McCool, D.K. and Yoder, D.C. 1997. Predicting Soil Erosion by Water: a Guide to Conservation Planning with the Revised Universal Soil Loss Equation (RUSLE). USDA-ARS, Handbook 703, USDA, Agricultural Research Service, Washington, DC., 384 pp.

Spaeth, K.E. , Thurow, T.L., Blackburn, W.H. and Pierson, F.B. 1996. Ecological dynamics and management effects on rangeland hydrologic processes. In: Grazingland Hydrology

Issues: Perspectives for the 21st Century. Eds. K.E. Spaeth, F.B. Pierson, M.A. Weltz and G. Hendricks. pp. 25–51. Society for Range Management, Denver, CO.

Seyfried, M.S. and Flerchinger, G.N. 1994. Influence of frozen soil on rangeland erosion. In: Variability of Rangeland Water Erosion Processes. Eds. W.H. Blackburn, F.B. Pierson, G.E. Schuman and R.E. Zartman. pp. 67–82. Soil Sci. Soc. Am. Special Publication 38. Soil Science Society of America, Madison, WI.

Thurow, T.L. 1996. Rangeland watershed research and technology needs for the future. In: Grazingland Hydrology Issues: Perspectives for the 21st Century. Eds. K.E. Spaeth, F.B. Pierson, M.A. Weltz and G. Hendricks. pp. 125–136. Society for Range Management, Denver, CO.

Wagenet, R.J. 1988. Modeling soil hydrology: perspectives, perils and directions. In: Modeling Agricultural, Forest, and Rangeland Hydrology. pp. 1–9. ASAE Pub. 07-88, American Society of Agricultural Engineers, St. Joseph, MI.

Weltz, M.A., Fox, H.D., Amer, S., Pierson, F.B. and Lane, L.J. 1996. Erosion prediction on range and grazing lands: a current perspective. In: Grazingland Hydrology Issues: Perspectives for the 21st Century. Eds. K.E. Spaeth, F.B. Pierson, M.A. Weltz and G. Hendricks. pp. 97–116. Society for Range Management, Denver, CO.

Weltz, M.A., Kidwell, M.R. and Fox, H.D. 1998. Influence of abiotic and biotic factors in measuring and modeling soil erosion on rangelands: state of knowledge. J. Range Manage. 51, 482–495.

Wight, J.R. 1988. Potential impact of modeling and systems simulation in rangeland research and management. In: Achieving Efficient Use of Rangeland Resources. Eds. R.S. White and R.E. Short. pp. 107–110. Montana Agricultural Experiment Station, Bozeman, MT.

Wight, J.R. and Hanks, R.J. 1981. A water-balance climate model for range herbage production. J. Range Manage. 34, 307–311.

Wischmeier, W.H. 1976. Use and misuse of the universal soil loss equation. J. Soil Water Cons. 31, 5–7.

Desert rangelands, degradation and nutrients

Kris M. Havstad[1], J.E. Herrick[1] and W.H. Schlesinger[2]
[1] *USDA, Jornada Experimental Range, Dept. 3JER, NMSU, Las Cruces, NM 88003, USA*
[2] *Dept. of Botany, Duke University, Box 90340, Durham, NC 27708, USA*
Tel: 505 646 4842, Fax: 505 646 5889; E-mail: <khavstad@nmsu.edu>

ABSTRACT It is well recognized that rangeland environments, especially degraded environments, are relatively infertile. Though arid lands are moisture limited, it is often stated that they are nutrient regulated. A discussion of rangeland soil nutrients relative to landscape degradation should include: (1) dynamics of nutrient distributions; (2) soil properties affecting nutrient fluxes; (3) indicators of soil quality; and 4) nutrient based strategies for remediation. Nutrient distribution is strongly influenced by vegetation structure, and spatial nutrient patterns are closely linked to aspects of primary productivity and species composition. Effects of vegetation on nutrient spatial patterns can persist for decades, even if vegetation is altered by natural or anthropogenic disturbances. Though abiotic processes strongly shape physical features of arid landscapes, biotic activities directly affect soil nutrients. Unfortunately, assessment and monitoring technologies based on soil properties which quantify integrity of nutrient related processes have not been fully developed. Approaches for remediating degraded conditions need to exploit nutrient spatial and temporal heterogeneities. Remediation should be targeted to the most fertile sites, even if these are only relatively small parts of the overall landscape. Understanding nutrient dynamics is a key to triggering autogenic remediation of degraded rangelands.

Key words: desertification, desert grassland, nitrogen, nutrient dynamics, phosphorus.

1. INTRODUCTION

Rangeland degradation frequently occurs in desert environments throughout the world (Ludwig and Tongway, 1995; Arnalds, 1987). Although productivity in desert rangelands is one to three orders of magnitude lower than in forest ecosystems (Ludwig, 1987), desert rangelands cannot be viewed as either simple or unproductive systems. Productivity is strongly coupled to precipitation, but nutrients and their availability strongly regulate both primary and secondary production (Noy Meir, 1979/80).

Degradation of desert grasslands is often characterized by replacement of perennial herbaceous species by long-lived woody shrubs and by an associated reduction in the capacity of the ecosystem to perform some functions. These reductions, such as the capacity to capture and retain water from high-intensity precipitation events (Ludwig and Tongway, this volume; Thurow, this volume), may be irreversible on human timescales.

Numerous causes have been proposed for this grassland-to-shrubland transition including overgrazing (Conley et al., 1992), exclusion of fire (Brown and Minnich, 1986), dispersal of seeds of woody species by herbivores (Buffington and Herbel, 1965) and a combination of natural and anthropogenic stressors (Hastings and Turner, 1965; Archer et al., 1995). It is not likely that the historical effects of any one factor (stressor) can be clearly and singularly associated with desert grassland degradation. We can not necessarily differentiate between cases in which certain stressor(s) are directly responsible for changes and cases in which they simply facilitate the transition from a community that evolved under a previous climatic regime to one that is better adapted to current conditions (Tranquillini, 1979; Neilson, 1986). However, the effects of degradation are widely recognized. In addition, it is now understood that these changes can be long-lived, non-linear and resistant to remediation (Herrick et al., 1997; Archer and Stokes, this volume; Tongway and Hindley, this volume).

The dynamics of desert rangeland degradation have been conceptualized for the Chihuahuan Desert of North America (Fig. 1). Allogenic forces contributed to reductions in herbaceous cover and to increased dispersal of seeds of competitive woody species, which rapidly established under favorable conditions including reduced competition from the perennial grasses (Grover and Musick, 1990). Autogenic forces then reinforced spatial redistribution of limiting resources into larger, shrub-associated patches (Schlesinger et al., 1996). These dynamics parallel degradation processes described for other grazed terrestrial systems (van de Koppel et al., 1997; Ludwig and Tongway, this volume).

2. NUTRIENT DYNAMICS

The conceptual model of degradation in Fig. 1 focuses on nutrient responses to vegetation dynamics. It is well recognized that even desert grasslands in excellent condition are nutrient-poor environments and that several important soil nutrients including N, P, K and Zn are commonly deficient (Dregne, 1976). These deficiencies are compounded by spatial (both vertical and horizontal) and temporal variability, which further limit nutrient availability to plants during periods when water is relatively non-limiting (Virginia and Jarrell, 1983). The relative importance of these deficiencies depends on both inter- and intra-specific differences, and on the degree and effectiveness of mycorrhizal infection.

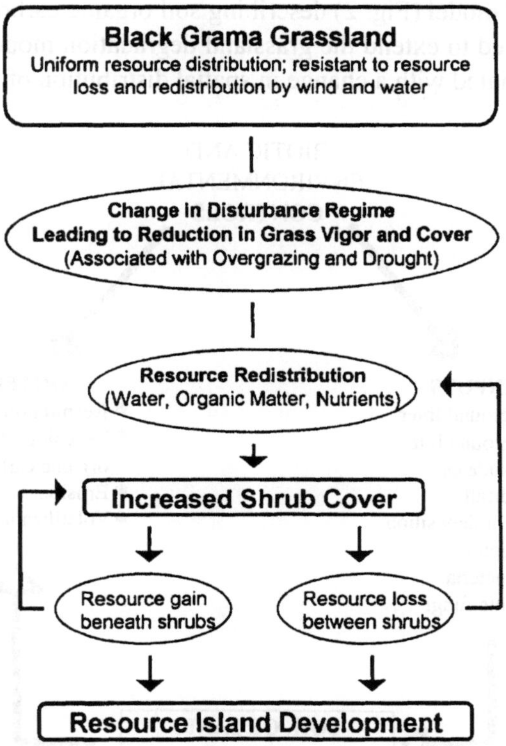

Figure 1. A central hypothesis of the research at the Jornada Long-term Ecological Research site in southern New Mexico is that changes in vegetation are accompanied by a redistribution of water and soil nutrient resources on the landscape, and this redistribution acts as a positive feedback mechanism to further promote degradation processes.

Nutrient-based processes in deserts are poorly understood (Zaady et al., 1996) due to both the diversity and complexity of these ecosystems, and the limited resources available to study them. Existing information, however, suggests that the distribution and dynamics of organic matter may be used as a general framework. Organic matter serves as an important source and sink for nitrogen and other nutrients in nearly all soils (Broadbent, 1986; Mengel and Kirkby, 1987), and is particularly important in degraded and sandy soils in which exchange surfaces and inorganic sources of nutrients are limited (Herrick and Wander, 1998). In non- or little-fertilized systems, plant N, P, S and microelement nutrition can depend heavily on mineralization of organic matter and residues of plants and animals (Chen and Stevenson, 1986; Broadbent, 1986). The most dynamic or rapidly cycled soil organic matter fraction is the biologically active fraction, which is tied to mineralization and therefore soil nutrient supply (Greenland and Ford, 1964).

A conceptual model (Fig. 2) describing soil organic carbon (Sollins et al., 1996) may be used to extend the grassland degradation model (Fig. 1). Degradation is associated with a change in spatial distribution of soil OM inputs.

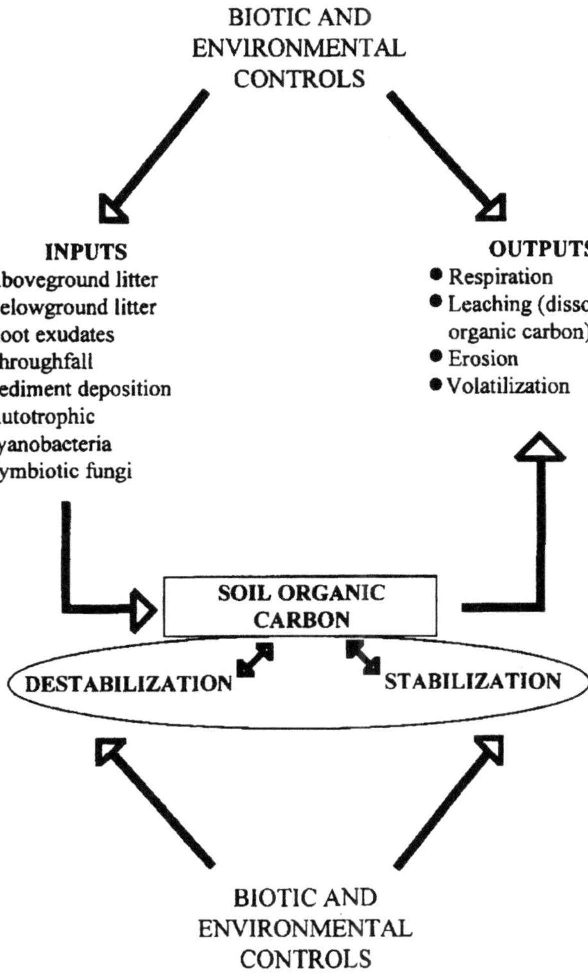

Figure 2. Organic C in soils results from long-term processes that control various C inputs and outputs. Destabilization processes increase the outputs. Adapted from Sollins et al., 1996.

In the Chihuahuan Desert, Connin et al. (1997) demonstrated a relatively rapid change in the form and vertical distribution of organic matter inputs to the soil profile but little change in total organic carbon following conversion of a grassland to a mesquite (*Prosopis glandulosa*) shrubland. For example, root diameters of >3 mm were absent from the grassland, and few roots were encountered below a depth of 40 cm, whereas the degraded shrub-dominated systems had significant quantities of large diameter roots throughout the profile. The quality of the organic matter inputs also changed. A C:N ratio of 38 was reported for grassland litter, while in the shrubland the ratio dropped to 16. Litter is recognized as an important regulator of the rate and fate of mineralized N in deserts (Zaady et al., 1996). These changes could dramatically affect C storage, N mineralization and OM turnover rates (Connin et al., 1997). These effects on C:N ratios and litter inputs are similar to those reported for grazing responses in more mesic grassland environments (Frank et al., 1995; Shariff et al., 1994).

The distribution of N, P and K has been shown to be strongly associated with the presence of shrubs in deserts (Schlesinger et al., 1996; Schlesinger and Pilmanis, 1998). Abrams and Jarrell (unpubl. data) found that the intensity of spatial variability of soil nutrients in the top 30 cm (Table 1), and the degree of association with the organic matter-rich soils under shrubs, increased with time along a shrub invasion chronosequence. Other studies have demonstrated similar spatial patterns in the activity of soil fauna which contribute to nutrient cycling (e.g., Santos et al., 1978; Kieft et al., 1998).

Table 1. Mean (x) soil nutrient concentrations in top 30 cm and coefficients of variation (CV) for three nutrients across a degradation chronosequence. The chronosequence occurred over a linear distance of 400 m where a gradual gradation from desert grassland to mesquite duneland was documented in 1937. The original grassland is intact at the start of the gradation, but mesquite dunelands have advanced in the ensuing 60 years, and this encroachment has been documented (Abrams and Jarrell, unpublished data).

	P (mg kg^{-1})		K (cmol kg^{-1})		Zn (mg kg^{-1})	
Degradation chronosequence (yrs)	x	CV	x	CV	x	CV
0 (Undegraded desert grassland)	1.58b*	39.91	0.44c	24.60	0.34a	17.74
10	4.07a	50.19	0.47b	18.97	0.30b	23.34
35	1.02c	39.92	0.41d	18.04	0.22c	36.61
45	1.65b	69.24	0.49b	21.70	0.23c	37.43
60 (Degraded mesquite duneland)	1.57b	54.19	0.58a	27.89	0.18d	40.50

*Means in the same column followed by the same letter do not differ ($P<0.05$).

Based on measurements of erosion rates in these grassland-mesquite transitions (Table 2; Gibbens et al., 1983), this increase in variability should be associated with a net reduction in both organic matter and nutrients. A textural analysis of soils (Table 2) further indicated that silt-sized particles were preferentially removed from mesquite-invaded parts of the "grid stakes" site (Hennessey et al., 1986). However, the measurements of Abrams and Jarrell, made at the "transect stakes" site (Table 2) provided little indication of nutrient loss from the system as a whole with degradation. The one exception was a significant decrease in Zn availability with degradation.

Table 2. Deposition and deflation of soil in 1935 and 1980 at 105 grid stakes on the 259-ha natural revegetation exclosure where soil levels were marked in 1933, and 1980 soil levels at 113 transect stakes on the exclosure on which soil levels were marked in 1935. From Gibbens et al., 1983.

Database	Year of measurement	Soil movement category	Number of points	Maxi mum (cm)	Minimum (cm)	Mean (cm)	Net loss (–) or gain (+) (cm)
Grid stakes	1935	No change	9				
		Deposition	36	6.0	0	1.1	
		Deflation	60	4.9	0	1.4	–0.4
	1980	Deposition	33	78.3	1.8	23.8	
		Deflation	72	61.9*	0.9	17.4	–4.6*
Transect stakes	1980	Deposition	43	78.6	0.6	2.5	
		Deflation	70	45.1*	0.9	2.1	–3.5*

*Represent minimum values because one stake was completely excavated by wind erosion.

Net primary productivity data from the Chihuahuan Desert provides some further indirect evidence for an increasing nutrient spatial heterogeneity in response to degradation. Huenneke, Anderson, Muldavin and Schlesinger (unpublished data) reported on aboveground biomass production for grassland and shrubland systems measured during three seasons for each of four years. Overall mean values of net primary production did not vary among these systems even though shrubland systems represent severely degraded conditions in this environment (Schlesinger et al., 1990). Systems did significantly vary in seasons of peak production and in the spatial variability of aboveground production. Shrublands displayed significantly greater spatial heterogeneity of production.

3. NUTRIENT DYNAMICS AND REMEDIATION

The overall importance of spatial patterning in deserts is well recognized and increasingly well described (Stafford-Smith and Morton, 1990; Tongway and Hindley, this volume). Production is highly patterned (and boundaries can be very abrupt), water is intermittently available, and nutrients are concentrated within fertile pockets in many systems. Remediation of degraded landscapes needs to recognize and exploit the opportunities created by spatial patterns (Whisenant and Tongway, 1996) and the episodic conditions that created these patterns (Winkel et al., 1991). Remediation technologies developed for average and homogenous environments will be ineffective, unsuccessful, costly and unadopted. Whisenant and Tongway (1996) argued that effective mitigation of degradation requires an intense attention to ecological processes. Call and Roundy (1991) have suggested that "safe" sites where germination and establishment can occur must be identified and exploited for arid and semiarid land reclamation. Herrick et al. (1997) have proposed a conceptual strategy for remediation that extends this concept of "safe" sites to recognize relatively fertile sites that occur within degraded arid landscapes. A basic premise of this model is that initial sites for remediation not only have greater ecological potential than surrounding sites, but will inherently facilitate migration from the site (Fig. 3). Conceptual models which may be used to help identify and select these fertile "trigger" sites include the Desertification Response Unit approach (Imeson et al., 1996; Imeson and Cammeraat, this volume) and the Australian pulse-reserve (Ludwig and Tongway, this volume). Both have nutrient dynamics embedded within them.

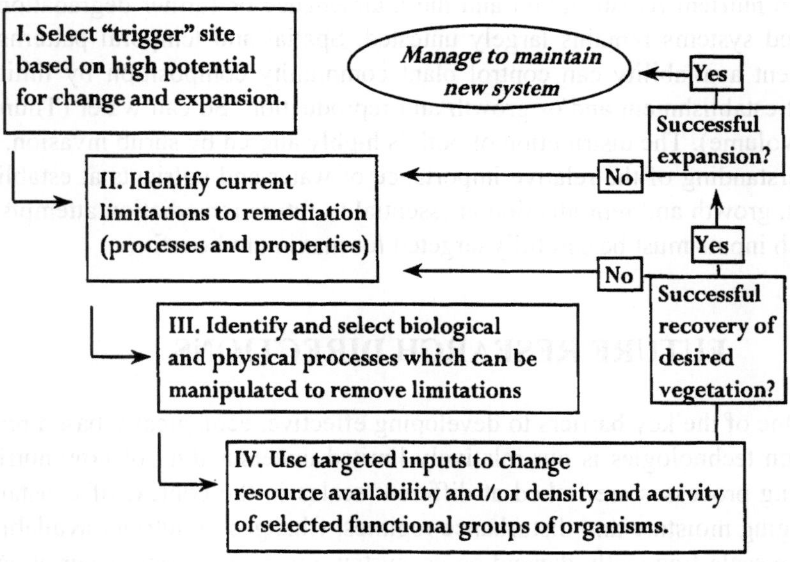

Figure 3. Conceptual model for development and application of an ecologically-based approach to rangeland remediation. Adapted from Herrick et al., 1997.

4. NUTRIENT PATTERNS: INDICATORS AND/OR DRIVERS OF DEGRADATION

The complexity of ecosystem processes, their interactions and their dynamics in response to degradation may confound attempts to identify simple nutrient-vegetation associations across vegetation states. However, the connection between changes in the spatial distribution of nutrients and the degradation of desert grasslands is well documented, and it is clear that most arid and semi-arid ecosystems are nutrient-limited when water is abundant and temperatures are appropriate for plant growth. The spatial distribution of nutrient resources may, in fact, serve as a valuable indicator of soil quality (Schlesinger et al., 1996; Herrick and Wander, 1997; Tongway and Hindley, this volume). Some parameters or indices may even serve as early warning indicators of degradation risk. For example, an increase in the coefficient of variation or ratio of soil organic matter concentrations in bare spaces vs. below plant canopies may be an indicator of reduced resilience of a site to recover from degradation.

While the correlation between changes in nutrient patterns and degradation makes this a potentially valuable indicator, the causal relationship between nutrient redistribution and the maintenance or further degradation of altered systems remains largely untested. Spatial and temporal patterns of nutrient availability can control plant community composition by limiting plant establishment and/or growth and reproduction. So can water (Thurow, this volume). The distribution of both is highly altered by shrub invasion. An understanding of the relative importance of water and nutrients at establishment, growth and reproduction is essential to future remediation attempts, in which inputs must be carefully targeted for maximum benefit.

5. FUTURE RESEARCH DIRECTIONS

One of the key barriers to developing effective, ecologically-based remediation technologies is our relatively limited understanding of how nutrient cycling processes are linked at different scales in the context of constantly changing moisture and disturbance regimes. Changes in nutrient availability at one scale frequently depend on ecosystem processes which occur at completely different scales. The spatially and temporally dynamic nature of nutrient cycling in deserts should provide tremendous opportunities to effect change at larger scales by understanding and targeting smaller scale processes. These opportunities will remain unexploited, however, unless significant advances are made in our understanding of processes occurring at the level of the rhizosphere and even the "hyphalsphere", how these processes are affected by moisture and disturbance, and how they affect, and are affected by water and nutrient resource transfers at scales from plant to the landscape. Historically, most studies have been completed at the plot level, or in the laboratory using disturbed, homogenized samples under relatively controlled moisture regimes. Future studies must build on current knowledge of spatial patterns in arid landscapes by quantifying inputs and losses at the patch scale which functions in both patch- and landscape-level processes (see also Ludwig and Tongway, this volume; Tongway and Hindley, this volume).

REFERENCES

Archer, S., Schimal, D.S. and Holland, E.A. 1995. Mechanisms of shrubland expansion: land use, climate, or carbon dioxide. Clim. Change 29, 91–99.

Arnalds, A. 1987. Ecosystem disturbance in Iceland. Arctic Alpine Res. 19, 508–513.

Broadbent, F,E. 1986. Nitrogen and phosphorus supply to plants by organic matter and their transformations. In: The Role of Organic Matter in Modern Agriculture. Eds. Y. Avinmelch and Y. Chen. pp. 13–23. Martinus Nijhoff, Dordrecht.

Brown, D.E. and Minnich, R.A. 1986. Fire and changes in creosote bush scrub of the western Sonoran Desert, California. America Midland Naturalist 116, 411–422.

Buffington, L.C. and Herbel, C.H. 1965. Vegetational changes on a semidesert grassland range from 1858 to 1963. Ecol. Monogr. 35, 139–164.

Call, C.A. and Roundy, B.A. 1991. Perspectives and processes in revegetation of arid and semiarid rangelands. J. Range Manage. 44, 543–549.

Chen, Y. and Stevenson, F.J. 1986. Soil organic matter interactions with trace metals. In: The Role of Organic Matter in Modern Agriculture. Eds. Y. Chen and Y. Avnimelech. pp. 73–116. Martinus Nijhoff, Dordrecht.

Conley, W., Conley, M.R. and Karl, T.R. 1992. A computational study of episodic events and historical context in long-term ecological processes: climate and grazing in the Northern Chihuahuan Desert. Coenoses 7, 1–19.

Connin, S.L., Virginia, R.A. and Chamberlain, C.P. 1997. Carbon isotopes reveal changes in soil organic matter production and turnover following arid land shrub expansion. Oecologia 110, 374–386.

Dregne, H.E. 1976. Soils of Arid Regions. Elsevier Scientific Publishing Company, New York, NY, 237 pp.

Frank, A.B., Tanaka, D.L., Hoffman, L. and Follett, R.F. 1995. Soil carbon and nitrogen of Northern Great Plains grasslands as influenced by long-term grazing. J. Range Manage. 48, 470–474.

Gibbens, R.P., Tromble, J.M., Hennessy, J.T. and Cardenas, M. 1983. Soil movement in mesquite dunelands and former grasslands of southern New Mexico from 1933 to 1980. J. Range Manage. 36, 145–148.

Greenland, D.J. and Ford, G.W. 1964. Separation of partially humified organic materials from soils by ultrasonic dispersion. In: Transactions of the 8th International Congress of Soil Science. pp. 137–148.

Grover, H.D. and Musick, H.B. 1990. Shrubland encroachment in southern New Mexico, USA: an analysis of desertification processes in the American Southwest. Clim. Change 17, 305–330.

Hastings, J.R. and Turner, R.M. 1965. The Changing Mile. Univer. of Ariz. Press, Tucson, Arizona.

Hennessy, J.T., Kies, B., Gibbens, R.P. and Tromble, J.M. 1986. Soil sorting by fourty-five years of wind erosion on a southern New Mexico range. Soil Sci. Soc. Am. J. 50, 391–394.

Herrick, J.E., Havstad, K.M. and Coffin, D.P. 1997. Rethinking remediation technologies for desertified landscapes. J. Soil Water Cons. 52, 220–225.
Herrick, J. and Wander, M.M. 1998. Relationships between soil organic carbon and soil quality in cropped and rangeland soils: the importance of distribution, composition and soil biological activity. In: Advances in Soil Science: Soil Processes and the Carbon Cycle. Eds. R. Lal, J. Kimble, R. Follett and B.A. Stewart. pp. 405–425. CRC Press LLC, Boca Raton, FL.
Kieft, T.L., White, C.S., Loftin, S.R., Aguilar, R. Craig, J.A. and Skaar, D.A. 1998. Temporal dynamics in soil carbon and nitrogen resources at a grassland-shrubland ecotone. Ecology 79, 671–683.
Imeson, A.C., Perez-Trejo, F. and Cammeraat, L.H. 1996. The response of landscape units to desertification. In: Mediterranean Land Use and Desertification. Eds. C.J. Brandt and J.B. Thornes. pp. 447–469. John Wiley and Sons, Ltd.
Ludwig, J.A. 1987. Primary productivity in arid lands: myths and realities. J. Arid Environ.13, 1–7.
Ludwig, J.A. and Tongway, D.J. 1995. Desertification in Australia: an eye to grass roots and landscapes. Environ. Monit. Assess. 37, 231–237.
Ludwig, J., Tongway, D., Freudenberger, D., Noble, J. and Hodgkinson, K. 1997. Landscape Ecology, Function and Management: Principles from Australia's Rangelands. CSIRO Publishing, Collingwood, 158 pp.
Mengel, K. and Kirkby, E.A. 1987. Principles of Plant Nutrition. International Potash Institute, Bern, Switzerland.
Neilson, R.P. 1986. High-resolution climatic analysis and Southwest biogeography. Science 232, 27–34.
Noy-Meir, I. 1979/80. Structure and function of desert ecosystems. Israel J. Bot. 28, 1–19.
Santos, P.F., Depree, E. and Whitford, W.G. 1978. Spatial distribution of litter and microarthropods in a Chihuahuan Desert ecosystem. J. Arid Environ. 1, 41–48.
Schlesinger, W.H., Reynolds, J.F., Cunningham, G.L., Huenneke, L.F., Jarrell, W.M., Virginia, R.A. and Whitford, W.G. 1990. Biological feedbacks in global desertification. Sci. 247, 1043–1048.
Schlesinger, W.H., Raikes, J.A., Hartley, A.E. and Cross, A.F. 1996. On the spatial pattern of soil nutrients in desert ecosystems. Ecol. 77, 364–374.
Schlesinger, W.H. and Pilmains, A.M. 1998. Plant-soil interactions in deserts. Biogeochem. 42, 169–187.
Shariff, A.R., Biodini, M.E. and Grygiel, C.E. 1994. Grazing intensity effects on litter decomposition and soil nitrogen mineralization. J. Range Manage. 47, 444–449.
Sollins, P., Homann, P. and Caldwell, B.A. 1996. Stabilization and destabilization of soil organic matter: mechanisms and controls. Geoderma 74, 65–105.
Stafford-Smith, D.M. and Morton, S.R. 1990. A framework for the ecology of arid Australia. J. Arid Environ. 18, 255–278.
Tranquillini, W. 1979. Physiological Ecology of the Alpine Timberline: Tree Existence at High Altitude with Special Reference to the European Alps. Ecological Studies vol 31. Springer Verlag, New York, NY, 131 pp.
van de Koppel, J., Rietkerk, M. and Weissing, F.J. 1997. Catastrophic vegetation shifts and soil degradation in terrestrial grazing systems. TREE 12, 352–356.
Virginia, R.A. and Jarrell, W.M. 1983. Soil properties in a mesquite-dominated Sonoran desert ecosystem. Soil Sci. Soc. Am. J. 47, 138–144.
Whisenant, S.G. and Tongway, D.J. 1996 Initiating autogenic restoration on degraded arid lands. In: Proceedings of the Fifth International Rangeland Congress. Ed. N.E. West. pp. 62–64. Soc. for Range Manage., Denver, Colorado.

Winkel Von K. Roundy, B.A. and Cox, J.R. 1991. Influence of seedbed microsite characteristics on grass seedling emergence. J. Range Manage. 44: 210–214.

Zaady, E., Groffman, P.M. and Shachak, M. 1996. Litter as a regulator of N and C dynamics in macrophytic patches in the Negev desert soils. Soil Biol. Biochem. 28, 39–46.

Assessing and monitoring desertification with soil indicators

David Tongway and Norman Hindley
CSIRO Wildlife and Ecology, P.O. Box 284, Canberra City, ACT. 2601, Australia
Tel: 61 2 6242 1641; Fax 61 2 6242 1565; E-mail: <david.tongway@dwe.csiro.au>

ABSTRACT A stepwise monitoring procedure based on the Trigger-Transfer-Reserve-Pulse framework of landscape function determination is proposed as a method to assess desertification in rangelands. The procedure is entirely field-based, using indicators of surface processes to infer the status of landscape on a function/dysfunction continuum. The first step is to identify, map and measure zones of resource loss (run-off) and resource gain (run-on) in the landscape. In the second step, each zone type is assessed, using 11 soil surface indicators reflecting the effect of surface processes on soil as plant habitat. The indicators are grouped into three indices; stability (manifest as resistance to erosion), infiltration/water storage and nutrient cycling. Trends over time, or in response to specific environmental events can be plotted. A framework for the interpretation of the data is suggested to assist management decisions. The method facilitates the distinction between mere utilisation of pasture resources and desertification, which by our definition is loss of function. The method is repeatable and consistent between observers. The approach was derived from a wide range of Australian rangelands and has been tested in South Africa and USA for broad application. Recent use to assess minesite rehabilitation shows that the approach can accommodate a very wide range of function/dysfunction scenarios.

Key words: conceptual framework, field-based, indicators, landscape function, soil.

1. INTRODUCTION

Detection and monitoring of desertification in rangelands has typically been descriptive, often restricted to evidence provided by a narrow range of biota or associated with theories of plant succession (e.g., Golley, 1977). Desertification is also often expressed purely in terms of the consequences for particular industries, such as pastoralism, rather than in terms of the basic functioning of the ecosystem itself. Predictions about future trajectories of desertified landscapes have been based on past performance continuing and not upon predictive frameworks or models derived from landscape function principles.

The Trigger-Transfer-Reserve-Pulse (TTRP) framework (Ludwig and Tongway, 1997, & this volume) provides a conceptual framework upon which to build a methodology for assessing and monitoring desertification (Fig. 1). This framework encapsulates ecosystem functional processes by assessing the dynamic spatial distribution of vital resources. These processes include mobilisation, transport, accumulation, deposition, cycling and consumption. It interprets plant response in terms of the interaction of climatic events and soil properties and recognises the importance of both biotic and abiotic feedback processes that stabilise overall system behaviour.

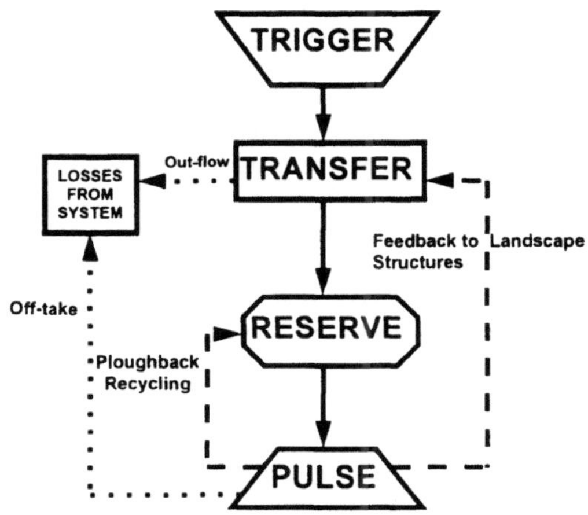

Figure 1. The Trigger-Transfer-Reserve-Pulse framework of landscape function. (After Ludwig and Tongway, 1997).

We have developed a field-based procedure that provides indicators of each of the processes identified in the TTRP as being important to the sustainability of substantially natural ecosystems. This procedure initially "isolates" and assesses the ecological status of land from the consequences for the industry or its people to enable substantially value-free advice to be given to planning authorities (Tongway and Ludwig, 1997b).

2. BACKGROUND

Studies of semi-arid lands with few overt signs of desertification showed that all were characterised by having zones of greater fertility, higher biomass of perennials, higher concentrations of soil nutrients and enhanced infiltration values interspersed with zones of low biomass and low fertility (Slatyer, 1961; Mabbutt and Fanning, 1987; Tongway and Ludwig, 1990;

Tongway, 1991). The zones formed a spatial arrangement or pattern in the landscape such that they contain functional units comprised of a run-on and an associated run-off zone. This need for spatial heterogeneity of water distribution to maximise production in deserts was predicted by Noy-Meir (1973). That is, there was a level of landscape organisation which facilitated adequate supplies of nutrients and water in these favoured locations (sink zones) by harvesting resources from "conjugate" source zones (Fig. 2). The vegetation itself was implicated in the resource regulation process by trapping and accumulating mobilised resources and cycling them (Tongway and Ludwig, 1997a).

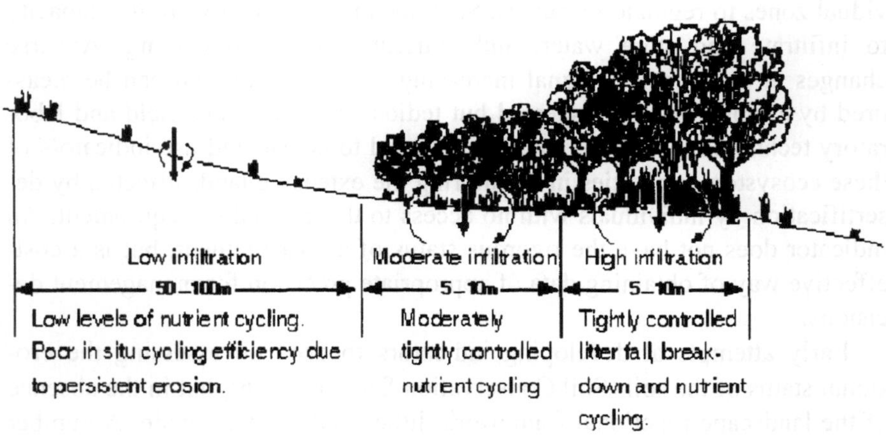

Figure 2. A highly patterned landscape type, showing functional landscape organisation in the form, of a grove-intergrove structure. The grove is comprised of an interception zone of grass and young shrubs and a zone of older trees. The intergrove is a sparse grassland. Water, topsoil and nutrients are shed from the sloping intergrove zone and absorbed in the grove system located on a flat. (After Tongway and Ludwig, 1990).

Clearly, in Fig. 2, a change in slope was strongly implicated in the resource transportation and depositional processes. Geomorphic features such as flats and depressions and plants singly or in thickets or swards act as resource trapping or accumulation agents. In chenopod shrublands where wind was the dominant resource mobilising agent, analogous zones of resource depletion (open flats) and accumulation (bush mounds) could be delineated (Tongway, 1991).

Landscapes affected by desertification or degradation, on analysis, were seen to have fewer or less effective resource accumulating zones and more or bigger or more effective resource transporting zones. Tongway and Ludwig (1997b) refer to this range of resource regulating capacity as the "function/dysfunction" continuum. Functionality is defined in terms of the degree to which resources tend to be retained, utilised and cycled in the ecosystem or lost by being washed or blown out of the system. This is the basis for the first part of the assessment procedure, called landscape function analysis (LFA), and is a landscape-scale indicator of resource regulation competence. The local watershed is the scale of the landscape studied.

3. SOIL SURFACE CONDITION ASSESSMENT

Within the initial discrimination of run-off and run-on zones in the landscape, there are additional, finer scale factors affecting the potential of individual zones to regulate resources. Such factors include erodability, capacity to infiltrate and store water, and nutrient status, and cycling. Adverse changes in these factors signal increasing desertification and can be measured by a range of well-respected but tedious and expensive field and laboratory techniques. There is, therefore, a need to devise and use indicators of these ecosystem properties in monitoring the extensive lands affected by desertification by individuals with no access to this specialised equipment. An indicator does not have the rigorous status of a measurement, but is a cost-effective way of obtaining data of appropriate precision for management decisions.

Early attempts at developing indicators focused on assessing the erosional status of the soil (Soil Conservation Service, 1976), but in the absence of the landscape functional framework, little headway was made. A number of subsequent studies examined a wide range of features expressed in the surface soil as candidate indicators of the "fitness" of soil to be a plant-growth medium (Smith, 1978; Mott et al., 1979; Tongway and Smith, 1989). These features had the status of "items on a list", rather than comprising an operational system.

The first step was made to place these indicators into a landscape context (Tongway and Ludwig, 1990) by observing that the occurrence of landscape spatial elements and surface features was correlated with measured properties such as infiltration rate, water stable aggregation, erosion and available N pools. Gradually, associations of features and properties built up (Greene and Tongway, 1989; Greene, 1993; Greene et al., 1994; Tongway and Smith, 1989). Eleven features which were widely distributed and generically informative were eventually selected (Table 1). This is a generic list of features to enable the method to be used at a variety of sites. Not all features exist at every site. The assessment of these features, called soil surface condition assessment, became the second part of the LFA procedure. Later studies of different landscape types used the complete procedure (Ludwig and Tongway, 1995).

Table 1. Soil surface features, with their respective process-based interpretations. See Tongway and Hindley (1995) for complete protocols for assessing these indicators.

	Indicator	Interpretation
1.	Soil cover	Assesses vulnerability to rainsplash erosion
2.	Basal cover of perennial grass	Assesses contribution of root biomass to nutrient cycling processes
3.	Litter cover, origin and degree of decomposition	Assesses the availability of surface organic matter for decomposition and nutrient cycling
4.	Cryptogam cover	An indicator of surface stability, resistance to erosion and nutrient availability
5.	Crust brokenness	Assesses loose crusted material available for wind ablation or water erosion
6.	Erosion features	Assesses the nature and severity of current soil erosion features
7.	Deposited materials	Assesses the quantity of alluvial deposits
8.	Microtopography	Assesses surface roughness for water infiltration and flow disruption, seed lodgement
9.	Surface resistance to erosion	Assesses likelihood of soil detachment and mobilisation by mechanical disturbance
10.	Slake test	Assesses soil stability/dispersiveness when wet
11.	Soil surface texture	An indicator of infiltration rate and water storage

The need for a coherent methodology by Agriculture Western Australia facilitated the development of a user-friendly LFA procedure (Tongway, 1994; Tongway and Hindley, 1995) for the routine monitoring of diverse rangelands in Western Australia. From field observations and measurements, the 11 soil surface indicators were distributed into three groups, to produce indices reflecting: (1) stability or resistance to erosion, (2) infiltration/water storage and (3) nutrient cycling. In use, the LFA procedure involves stratifying the landscape into zones of resource accumulation (sink) or resource loss (source) zones, using criteria and rules articulated in Tongway (1994) and Tongway and Hindley (1995). Examples of each zone type are then assessed for soil surface condition and the indices of stability, infiltration and nutrient cycling calculated. The status of individual landscape zone types can be monitored and reported, or a whole-of-site index can be calculated by multiplying the respective mean zone indices by the fractional proportion of the landscape occupied by the zones.

Together with the landscape stratification into zones, the soil surface feature analysis is the procedure advanced to both assess and monitor the process of desertification. Tests with newly trained observers shows that the method is quick to learn, enables rapid assessment and different observers obtain consistent results. Typically, data from a 100-m transect takes less than 1-hr to acquire. Application to diverse landscape types has been demonstrated, across a wide range on the function-dysfunction continuum. For example, minesites have been successfully monitored in biomes as diverse as semi-arid heathlands, temperate forests, tropical woodlands, chenopod shrublands and arid grasslands (Tongway et al., 1997).

The methodology explicitly involves a hierarchical set of scales from landsystem through landunit to landscape zone, to provide context relevance at coarse scale and insight detail at fine scale and encourages the user to explore these scales for various purposes. The data manipulation is a transparent process, facilitating up- and down-scale interpretations.

The methodology is focused on indicators of processes rather than of organisms, and has a predictive capacity. The possible consequences of scenarios, such as climatic events like wind, drought and rain-storms, fire, or changes to animal numbers on landscape function, may be examined. The soil surface features used in the methodology can respond rapidly to circumstances such as rare climatic events implying a change in edaphic habitat, whereas biota may respond more slowly. The indicators cover a wide range of habitat characteristics and are sufficiently versatile to be used across a range of biomes.

4. USE OF INDICATORS FOR ASSESSING REHABILITATION POTENTIAL

The data referred to in the preceding sections also needs an interpretation framework to be of practical use. For example, the indicators can be used to identify critical thresholds in landscape function. The State and Transition model (Westoby et al., 1989) recognised that different states may exist, but left the physical manifestation and distinction between states and the biophysical realities and triggers for the transition processes up to the observer to discover (Archer and Stokes, this volume).

Figs. 3 and 4 summarise a conceptual framework to facilitate the interpretation of the indices obtained in the landscape function analysis in terms of functionally critical states. The figures represent respectively a robust and a fragile landscape type. The curves represent the trajectory of the resource regulation capacity (as inferred from the indicators) of these landscape types with respect to the stresses and disturbances that bring about desertification. They are "reverse S" shaped: initially, the effect of desertification forces is small, followed by a sharp downward trend as desertification increases, then a final stage where there is little further change with increasing stress/disturbance (for examples, see Figs. 3–5, Archer and Stokes, this volume).

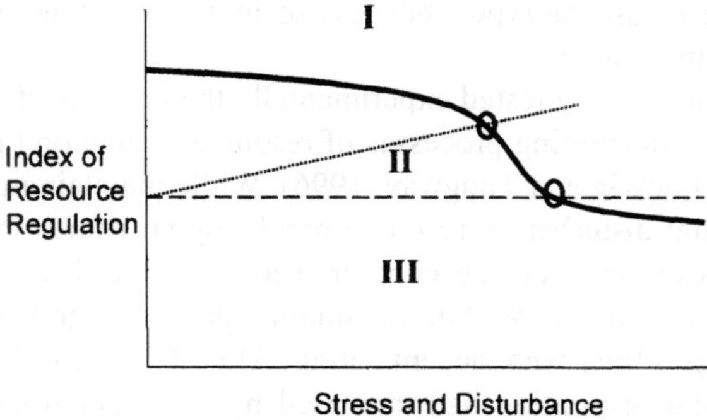

Figure 3. An example of a robust landscape type. The initial response to stress and/or disturbance has little effect on resource retention, but ultimately there is a distinct drop, followed by a slower decline. States I, II and III are discussed in the text.

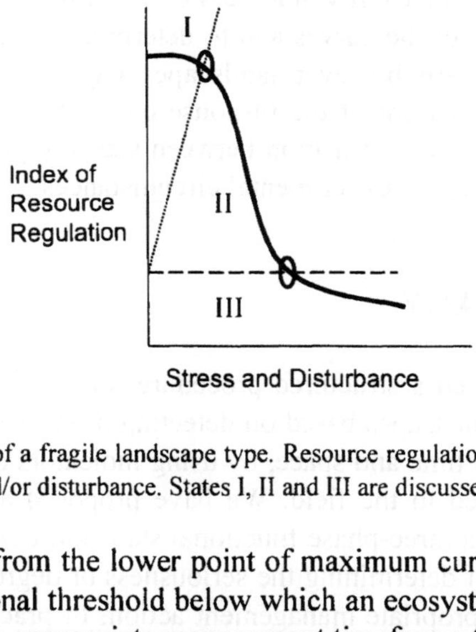

Figure 4. An example of a fragile landscape type. Resource regulation deteriorates markedly with only low stress and/or disturbance. States I, II and III are discussed in the text.

A line running from the lower point of maximum curvature to the y-axis represents a functional threshold below which an ecosystem will not recover spontaneously in an appropriate management timeframe (functional state III) even if all stresses are removed. A line running from the intersection of the first line with the y-axis to the point of upper maximum curvature defines two other functional states: state I where the system is self-regulating (with respect to vital resources); and a "meta-stable" state II, where desertification has occurred but may recover if stress and disturbance is lowered sufficiently. The upper line is defined as the "horizon of resource regulation sufficiency". State II has two management options: to remove stress/disturbance until the ecosystem crosses the horizon of resource regulation sufficiency, or to provide a means or mechanism to improve resource capture and retention. Note that a substantial improvement in resource regulation is needed in desertified fragile landscape types, but less so in a robust type, with obvious management ramifications.

Tongway and Ludwig tested experimentally the concept of rehabilitating a landscape by re-instituting processes of resource regulation (Tongway and Ludwig, 1996; Ludwig and Tongway, 1996). While maintaining a high level of grazing animal disturbance, resource regulating structures in the form of branch mounds oriented on the contour were introduced into a landscape assessed to be in state II. Within 18 months, perennial grasses had established and soil and litter were accumulating. After 3 years, soil N and C levels and microbial respiration had increased by 50% and water infiltration was an order of magnitude greater than controls. Ten years after establishment, including 4 years of drought, the treatment is still effective (Ludwig and Tongway, pers. observation).

This conceptual framework therefore defines 3 functional states and two critical thresholds by distinguishing differences in the capacity of the landscape to regulate resources. It would be the role of monitoring to empirically determine the shape of the curves and to determine the critical threshold indicator values. Typically however, landscapes may well be found either near the top or near the bottom of the response curve due to changes that took place some time ago. The transition between states might well be rapid and rarely observed, except in experimental circumstances.

5. SUMMARY

We have described a structured procedure for assessing and monitoring desertification in landscapes based on detecting differences in the regulation of vital resources in time and space, by using indicators of soil surface processes easily collected in the field. We have proposed a conceptual framework comprised of a three-phase functional state and transition model to assist in the process of determining the seriousness or degree of desertification and deciding on appropriate management action. In practice, empirical data would be fitted to this framework to derive critical indicator values. The procedure has been used in a wide range of landscape types and across rainfall regimes from 200 to 1500 mm and in assessing rehabilitation on minesites where the process of desertification is seen in reverse. There is both scope and diversity in the data to propose appropriate procedures for rehabilitation.

REFERENCES

Golley, F.B. 1977. Ecological Succession. Dowden, Hutchinson & Ross, Stroudsburg.

Greene, R.S.B. 1993. Infiltration measurements in the semi-arid woodlands of eastern Australia – comparison of methods. In: Proceedings of the XVII International Grassland Congress. pp. 79–80. Palmerston North, Hamilton and Lincoln, New Zealand & Rockhampton, Australia. Tropical grassland Society of Australia, Brisbane/New Zealand Grassland Association, Auckland.

Greene, R.S.B., Kinnell, P.I.A. and Wood, J.T. 1994. Role of plant cover and stock trampling on runoff and soil erosion from semi-arid wooded rangelands. Australian Journal of Soil Research 32, 953–973.

Greene, R.S. and Tongway, D.J. 1989. The significance of (surface) physical and chemical properties in determining the soil surface condition of red earths in rangelands. Australian Journal of Soil Research 27, 213–225.

Ludwig, J.A. and Tongway, D.J. 1995. Spatial organisation of landscapes and its function in semi-arid woodlands, Australia. Landscape Ecology 10, 51–63.

Ludwig, J.A. and Tongway, D.J. 1996. Rehabilitation of semi-arid landscapes in Australia. II. Restoring vegetation patches. Restoration Ecology 4, 398–406.

Ludwig, J.A and Tongway, D.J. 1997. A landscape approach to rangeland ecology. In: Landscape Ecology Function and Management: Principles from Australia's Rangelands. Eds. J. Ludwig, D. Tongway, D. Freudenberger, J. Noble and K. Hodgkinson. pp. 1–12. CSIRO, Melbourne.

Mabbutt, J.A. and Fanning, P.C. 1987. Vegetation banding in arid Western Australia. Journal of Arid Environments 12, 889–912.

Mott. J.J., Bridge, B.J. and Arndt, W. 1979. Soil seals in tropical tall grass pastures of northern Australia. Australian Journal of Soil Research 30, 483–494.

Noy-Meir, I. 1973. Desert ecosystems: environment and producers. Annual Review of Ecology and Systematics 4, 25–51.

Soil Conservation Service 1976. National Range Handbook. U.S. Department of Agriculture, Soil Conservation Service, Washington.

Slatyer, R.O. 1961. Methodology of a water balance study conducted on a desert woodland (*Acacia aneura* F Muell.) community in central Australia. UNESCO Arid Zone Research 16, 15–26.

Smith, E.L 1978. A critical evaluation of the range condition concept. In: Proceedings of the First International Rangeland Congress. pp. 266–267. Denver.

Tongway, D.J. 1991. Functional analysis of degraded rangelands as a means of defining appropriate restoration techniques. In: Proceedings of the IVth International Rangeland Congress. pp. 166–168. Association Francaise de Pastoralisme, Montpellier, France.

Tongway, D.J. 1994. Rangeland Soil Condition Assessment Manual. CSIRO, Melbourne.

Tongway, D. and Hindley, N. 1995. Assessment of Soil Condition of Tropical Grasslands. CSIRO Division of Wild life and Ecology, Canberra.

Tongway, D., Hindley, N., Ludwig, J., Kearns, A. and Barnett, G. 1997. Early indicators of ecosystem rehabilitation on selected minesites. 22nd Annual Environmental Workshop. Minerals Council of Australia, Dickson, Australia.

Tongway, D.J. and Ludwig, J.A. 1990. Vegetation and soil patterning in semi-arid mulga lands of Eastern Australia. Australian Journal of Ecology 15, 23–34.

Tongway, D.J. and Ludwig, J.A. 1996. Rehabilitation of semi-arid landscapes in Australia. I. Restoring productive soil patches. Restoration Ecology 4, 388–397.

Tongway, D.J. and Ludwig, J.A. 1997a. The conservation of water and nutrients within landscapes. In: Landscape Ecology Function and Management: Principles from Australia's Rangelands. Eds. J. Ludwig, D. Tongway, D. Freudenberger, J. Noble and K. Hodgkinson. pp. 13–22. CSIRO, Melbourne.

Tongway, D.J. and Ludwig, J.A. 1997b. The nature of landscape dysfunction in rangelands. In: Landscape Ecology Function and Management: Principles from Australia's Rangelands. Eds. J. Ludwig, D. Tongway, D. Freudenberger, J. Noble and K. Hodgkinson. pp. 49–61. CSIRO, Melbourne.

Tongway, D.J. and Smith, E.L. 1989. Soil surface features as indicators of rangeland site productivity. Australian Rangeland Journal 11, 15–20.

Westoby, M., Walker, B. and Noy-Meir, I. 1989. Opportunistic management for rangelands not at equilibrium. Journal of Range Management 42, 266–274.

Scaling up from field measurements to large areas using the Desertification Response Unit and Indicator Approaches

Anton Imeson and Erik Cammeraat
Landscape and Environmental Research Group, University of Amsterdam, Nieuwe Prinsengracht 130, 1018 VZ Amsterdam, The Netherlands
Tel: 31 20 5257457; Fax: 31 20 5257431; E-mail: <a.c.imeson@frw.uva.nl>

ABSTRACT The conceptual basis used in scaling up from field measurements to larger areas combines the response unit approach with indicator concepts. Response units are characterised by hierarchically linked patterns, the typology and dynamics of which reflect key processes of water and sediment storage and transport. Ecological and physical indicators of ecosystem process and function were analysed and evaluated for desertification response units in the Guadalentín catchment. From this it was concluded that different indicators are required for different response units, according to the processes affecting soil and water conservation functions. It was also concluded that desertification indicators are scale dependent. Indicators of soil quality and ecosystem health are highly appropriate for studying desertification, but they can not be applied uniformly. Rather, they must be applied within the framework of a typology of land units within which the responses to change are similar.

Key words: desertification, erosion, hierarchy, indicators, infiltration, land degradation, response units, soil quality, SE Spain, upscaling.

1. INTRODUCTION

Desertification is often defined and experienced as the irreversible loss of the production and ecological functions. Generally agreed indicators of desertification include soil erosion, loss of biodiversity and a lower productivity. There is a strong resemblance between concepts and indicators of desertification and concepts of declining soil health or quality (Doran and Jones, 1996; Doran and Parkin, 1996; Harris et al., 1996; Karlen et al., 1997). The objective of this chapter is to describe how the application of concepts and indicators of soil quality and health (Karlen et al., 1997) might be combined with the desertification response unit methodology (Imeson et al., 1995, 1996). This application is being developed in the Guadalentín catchment area in the Murcia Province of Spain (Brandt and Thornes, 1996).

The desertification response unit method is used to scale up from point measurements to large areas and to explain the contrasted response of different areas to desertification at different scales (Imeson et al., 1995, 1996). It considers the changes that are taking place in the soil-water-vegetation patterns in the landscape system at different scales, resulting from the movement of water, sediment and nutrients. The changes in patterns and the effects of processes on the response units can be viewed as indicators of health or quality. Because the processes underlying the behaviour of the indicators have been investigated, the behaviour of the indicators can be related to measurements and some models (Boardman and Favis-Mortlock, 1998; Kirkby et al., 1996).

Scaling up point measurements is an important research objective required, for example, for the quantitative application of remote sensing to soil erosion. A particular difficulty is that soil erosion measurements are point measurements made at only a few locations. The sediment measured leaving an erosion plot, micro-catchment or drainage basin is reported as an average rate of soil loss but in reality 90 per cent of the sediment may come from a small proportion of the area (Wolman and Miller, 1960). The non-random spatial variability in soil erosion is matched by a temporal variability that is related to both hydro-meteorological events and the ecological succession or evolution of the system. Soil erosion models generally lack the feedback with the vegetation and soil that are needed for long-term studies (Pierson, this volume). Consequently, it is usually impossible to quantitatively combine remote sensing information with the results of most soil erosion models or measured rates of erosion. The process-based desertification response approach avoids these problems by focusing on changes in patterns that can be observed and monitored. Measured or monitored erosion is related to the changes that occur in visually observable patterns and not to measured sediment fluxes which are highly sensitive.

The response unit method shares a common dynamic systems philosophy with the landscape systems prespectives reviewed in other chapters of this volume (Ludwig and Tongway; Archer and Stokes; Tongway and Hindley). The main differences are in the application and emphasis. The response unit approach is more focused on the evolution of abiotic (soils and geomorphology) patterns and structure in relation to water, sediment and nutrient movement and storage and the development of vegetation patterns. Developed for the Mediterranean region of Southern Europe, which has a heterogeneous and fragmented land use and land cover, the response unit approach emphasizes spatial heterogeneity in soil depth and water holding capacity related to the development of vegetation patterns.

2. RESPONSE UNITS

In arid and semi-arid landscapes, scarce water and nutrient resources are non-uniformly distributed to periodically create relatively moist and dry subsystems. This occurs at a hierarchy of different scales, ranging from the subpatch scale around individual plants to the scale of catchments and hillslopes. Processes acting at different scales affect one another through feedbacks. Coarse scale interactions at the slope scale feedback to and influence the dynamics of the finer scale units (O'Neill et al., 1986) comprising the hillslope.

These transfers and feedbacks, involving the movement of water and the response of plants and animals, result in the production of patterns in the soil and vegetation that are characteristic of either specific landscape positions or disturbances. They also provide information about stability and resilience.

Should the aboveground vegetation be degraded, subsurface expressions of the vegetation pattern, in the form of soil structure, remain for some time in the soil. These enhance the ability of the vegetation to recover because they concentrate soil moisture above the critical levels required for vegetative regeneration, seed germination and root development. Studies on abandoned cultivated fields provide insight into the rates of soil structure generation and degradation. For soil micro-aggregates it takes several decades for conditions to approach those found on non-cultivated land; coarser aggregates respond to change more quickly (Cammeraat and Imeson, 1998).

Different vegetation mosaics tend to be characteristic of different landscape units, reflecting the contributions or losses of water. Patterns, therefore, express both the influence of external factors operating at coarser spatial scales than the plot itself and of incorporated finer-scale processes that are internal to the unit. Lavee et al. (1998), studied patterns along climatic gradients and found that characteristic patterns typify the transition from sub-humid to semi-arid and from semi-arid to arid conditions.

From the above, it can be concluded that spatial scale is extremely important for integrating the changes taking place on response units and also for understanding how these are linked. These linkages mean that erosion and runoff phenomenon can not be used indiscriminately as indicators of ecosystem functioning. In other words the meaning of "erosion" in terms of soil quality or ecosystem health is both scale and site specific (Karlen et al., 1997; Imeson et al., 1996). Erosion and the illusion of poor health or quality in one response unit may be the visual price paid for the system to function in healthy way at the coarser scale at which response units interact through the transfer of resources. In other words, at the scale of interest, different surfaces could be described as two separate systems one healthy and the other unhealthy. From the larger scale perspective, there is just one healthy semi-arid system that can function because of the transfer of resources from one internal subsystem to another (see also Ludwig and Tongway, this volume; Tongway and Hindley, this volume). An important difference between Mediterranean climate and semi-arid eco-geomorphological systems, described by Lavee et al. (1998), is that whereas the former has rainfall characteristics that enable vegetation to cover the whole surface, the semi-arid areas do not. Small scale erosion and runoff may simply be an adaptation of the system to drought and semi-arid conditions, and not intrinsically an indicator of poor health.

3. STUDY SITES IN THE GUADALENTÍN

The field area, located in the Alqueria region of the Guadalentín catchment in the Province of Murcia, SE Spain is described in detail elsewhere (Cammeraat and Imeson, 1998). The area considered (about 100 km^2) has a relief of about 300 m. The geology is dominated by limestone and marls from the Upper Cretaceous and Tertiary with more recent calcareous Holocene and Pleistocene alluvial deposits in depressions. The climate is semi-arid with an average annual rainfall of 300 mm yr^{-1}, falling mainly in the autumn and spring. The average minimum and maximum temperatures are 9.3 and 26.0°C respectively. Representative soil profiles for the limestone and marl areas are summarised in Table 1.

Table 1. Surface characteristics of the soils of the Cañada Cazorla site.

Position	Slope $CaCO_3$ (°)	Soil type (afterFAO, 1989)	Hori-zon	Depth (cm)	Land use [a)]	Texture Sand (%)	Silt (%)	Clay (%)	Org. C (%)	$CaCO_3$ (%)
Plateau	0–1	Petric	(Crust)	0–3	S (St)	26.6	60.1	13.4	2.3	53.9
		Calcisol	(Ah)	3–18		27.0	68.0	5.0	2.9	59.7
Plateau	0–1	Petric	(Crust)	0–2	F (C)	27.6	67.8	4.6	3.4	53.4
		Calcisol	(Ap1)	2–19		34.4	60.9	4.7	3.2	53.7
Up-slope	12–25	Eutric	(Crust)	0–3	S (St)	36.7	51.7	11.0	1.8	78.4
		Leptosol	(Ah)	3–12		21.1	62.7	16.2	1.8	78.5
Mid slope	12–3	Eutric	(Crust)	0–3	S (St)	38.4	52.6	7.9	1.6	71.2
		Leptosol	(Ah)	3–9		32.4	55.2	10.3	2.0	68.8
Down-slope	6–3	Haplic	(Crust)	0–2	A(Pl)	26.5	66.6	6.5	2.1	64.4
		Calcisol	(Ap1)	4–21		23.3	68.4	6.2	1.8	64.4
Valley	3–0	Calcaric	(Bw)	21->40		14.9	64.5	9.6	1.2	52.6
		Regosol	(Ap1)	0–20	F (C)	19.6	65.4	14.5	1.4	61.1
			(Ap2)	20–40		13.3	66.6	17.2	0.7	40.7

a) S: semi-natural, A: abandoned, F: fallow, St: *Stipa tenacissima* dominated, Pl: *Plantago albicans* dominated, C: rain fed cereals.

A land use survey was made of the study region in July 1997. The land use and cover are characterised by (1) irrigated areas producing mainly wheat, almonds, melons, peppers and forage; (2) areas of dryland agriculture located predominantly on the lower, more gentle marl areas, frequently down-slope of eroding areas that supply water following rainfall. This rainfall harvesting is quantitatively significant; (3) abandoned pastoral and cultivated areas. The development of a vegetation is usually very slow. The soils have very low biological activity (Cerdà et al., 1994, 1995), extensive biogenic crusts and low fertility; (4) alpha grass (*Stipa tenacissima*) which is an important cover type, particularly on the limestone areas where it has been cultivated since Roman times. It forms tussocks that vary in size and forms characteristic patterns on slopes that reflect the geology, aspect and transport of stones by grazing animals and hunters; (5) natural growth of Aleppo pine (*Pinus halepensis*) which covers large areas interspersed with alpha grass, particularly on limestone areas where water is concentrated by joints; and (6) large scale reforestation with Aleppo pine, which took place in the 1970s, accompanied by extensive terracing. The growth of the pines is variable, but often very poor.

Comparisons of air photos from 1958 and 1995 show that land use changes in the region have been considerable. In the Cañada Cazorla, for example, only 40 per cent of the area has the same land use as in 1958. In 1958 the area consisted of small fields in the valleys and areas that are now abandoned were used for grazing (Imeson et al., 1998).

Within the study area, four areas were selected for more detailed analysis. These include the Cañada Cazorla, the Cañada de Hermosa and the Buittre-Alquería area. A typical section across the Alquería site, shows the lithology, vegetation and soils (Figs. 1 and 2).

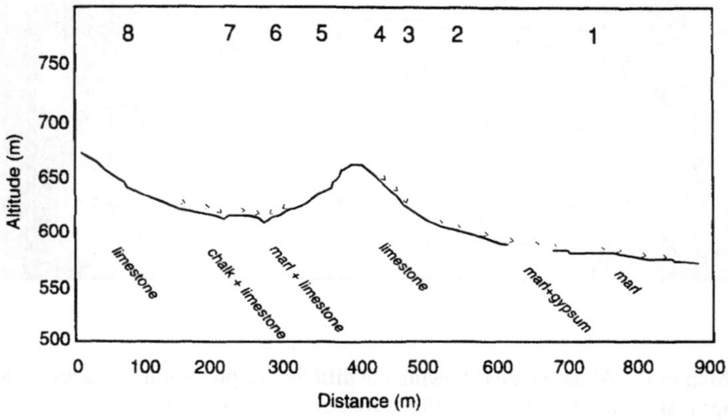

Figure 1. Section across the Alquería study area and location of process response units shown in Fig. 2

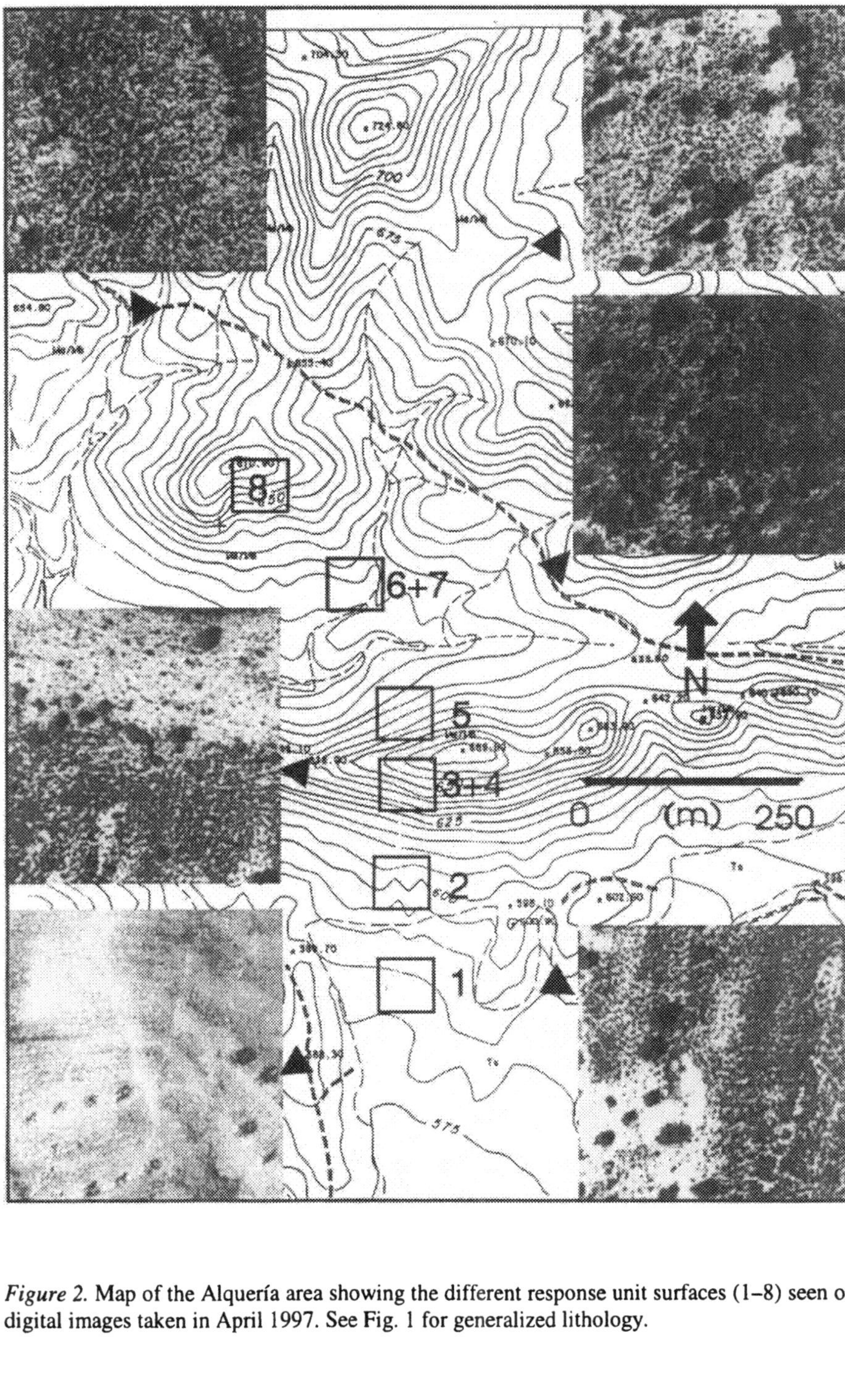

Figure 2. Map of the Alquería area showing the different response unit surfaces (1–8) seen on digital images taken in April 1997. See Fig. 1 for generalized lithology.

4. METHODS

Measurements of runoff are made in nested catchments at different scales. Open plots within micro-watersheds, within mini-catchments and catchments have been installed at Alquería and at the Buittre (Cammeraat and Imeson, 1999a). These measurements are being made for semi-natural forest, reforested and alpha grass vegetation types. An automatic weather station records meteorological data.

Rainfall simulation experiments were made at two scales (patch and response unit scale) to study the effect of different factors on water and sediment movement. Result to date (Cammeraat and Imeson, 1999b; Cerdà, 1997; Kirkby et al., 1996) demonstrate clear relationships between infiltration, soil structure and vegetation.

Detailed field surveys and air photo-interpretations were made to collect information on the distribution of soil profiles. Soils were sampled and soil characteristics, including aggregation and infiltration rates, determined. Digital, black and white and false colour low level air photos were obtained in April 1997 with 40–80 cm resolution. The response unit interpretation follows the methodology described by Imeson et al. (1996).

5. INDICATORS OF THE SOIL AND WATER CONSERVATION FUNCTIONS

Scaling-up from point measurements requires identification of indicators that condense or summarise the information that can be obtained from detailed monitoring. Indicators must be able to identify the undesirable changes that are expected to occur, or which have already taken place. A full discussion of the use, problems and potentials of ecological indicators is given by Landres (1996) and Harris et al. (1996). Ecological indicators are usually indicators of ecosystem structure or function (NRC, 1994; see also Tongway and Hindley, this volume). The summarised indicators shown in Table 2, relate to the physical and biological health of the ecosystems. Similar indicators could have been obtained from Doran and Parkin (1996). The values of the indicators applied have been adapted to the condition of the research area and are simplified for space. At this level of generalisation, they serve to describe how well the response units are performing the soil and water conservation functions and their utility for recognizing the health of the unit in terms of the categories, *healthy, at risk and unhealthy*.

6. RESULTS AND DISCUSSION

6.1 The Alquería reference site and the process response units

A section across one of the three representative sites (Alquería) is sketched in Fig. 1 to illustrate the juxtaposition of the process-response units. In Fig. 2, a contour map is shown along with examples of the digital remote sensing images taken in April 1997.

The response units shown are the result of a combination of land use and climate-driven transport processes. At the catena or hillslope scale, processes are affected by the impact of human activity on sediment transport and water movement along the slope. Boundaries between fields and land use types play a role. Grazing and tillage influence the availability and transport of sediment. Soil crusting, soil surface stoniness and sealing (soil degradation) strongly affect runoff. Units high on the slope can exert a strong influence on areas down slope. Within the response unit, characteristic patch-scale processes impart to the response unit its inherent characteristics. The units are, therefore, characterised by slope-scale processes that influence the characteristics of the site and patch-scale processes operating at finer scales.

6.1.1 Description of response units

The following units may be distinguished according to hydro-geomorphological processes:

- **Process response unit 1**. This unit comprises the lower, relatively gentle colluvial slopes on marls. It has an Ap surface horizon which is highly sensitive to crusting and has a very low infiltration rate. Three sub-units are shown. The most extensive unit (1A) consists of areas under cultivation; 1B is abandoned land; and 1C sub-units are linear depressions characterised by regular additions of water and seeds and hence by a better vegetation cover and more organic rich soils that the surrounding areas.

  The dominant patch-scale processes are influenced by ploughing; the transport of material brought to the surface by soil fauna, rainwash and grazing of stubble.
- **Process response unit 2**. The vegetation is dominated by *Stipa tenecisima*. The *Stipa* tussocks have Ah surface horizons beneath them on top of very shallow soils. Between the tussocks are stone pavements comprised of rocks supplied from unit 4. This supply is variable, but was particularly significant at the site shown. These areas can generate substantial local runoff during rainfall events and they concentrate this and the moving stones into chutes that form where the *Stipa* bands that hold back

stones are disrupted. Soil thickness is important and is reduced on south-facing slopes. The most important transport processes are concentrated overland flow and gravitational transport which are intensified by grazing. Puigdefabregas and Sanchez (1996) describe these dynamics in detail for an area on metamorphic rocks near Almeria (SE Spain). Local transport processes are greatly affected by the dynamics of the *Stipa*, and subunits could be recognised according to the slope angle and rock supply from up-slope areas which exert a strong impact on the spatial patterns of the *Stipa* tussocks.

- **Process response unit 3**. This unit is comprised of a band of shrubs on rocky soil that receives a large amount of runon and (probably) seepage from the unit above. The unit is located at the boundary between the almost bare limestone hillslope section above and the marls below. The soil containing rocks from unit 4, provides a humid micro-environment. The main processes here are similar to those described for unit 2.
- **Process response unit 4**. This area consist of a very rocky and steep limestone slope. The vegetation cover is sparse, consisting of annual grasses and small shrubs occupying positions along joints, in shallow soil pockets and between stones. Here water infiltrates into cracks, reappearing at cracks lower on the slope, but most water is running down as overlandflow to unit 3.
- **Process response unit 5**. On a north-facing slope this area is comprised of relatively open *Pinus halepensis* forest. The soil is formed by an Ah horizon resting on an C/R horizon. The soil aggregation near the surface is good and there is virtually no crust. There seems to be little overland flow, but there is a considerable down slope gravitational movement of stones and a lot of biological activity. An undercover of small shrubs (*Rosmarinus officinalis*, *Thymus vulgaris*), *Stipa tenacissima* and annuals is important.
- **Process response unit 6**. This open, degraded forest on inter-bedded limestone and marls has a crust and stone pavement. Soils are very shallow or absent. There are large non-vegetated areas and a moderate cover of *Stipa* and other perennials. Numerous small, shallow rills and bare areas generate local runoff. Some of the rills reach up into the forest unit above (unit 5). Open runoff plots have been installed in this unit.
- **Process response unit** 7. This is an area of degraded mattoral dominated by *Stipa*. It occupies a pediment developed on marls and limestone. There is an Ah horizon directly on the rocky pediment surface under the *Stipa* tussocks. The pediment is formed by a thick indurated calcrete layer which is exposed at many places at the surface. This has possibly restricted the growth of trees, although locally Pine trees and other shrubs are present. At open places where some topsoil is present, a strong crust

may occur, often with a stone pavement on top. This unit generates much runoff which is also measured on experimental plots.

- **Process response unit 8**. This is an area of mattoral vegetation, dominated by individual *Stipa* plants separated by bare crusted (structural) surfaces. There is much overland flow at the transition of unit 7 to 8 where the soil surface is strongly degraded and capped by strong, dark cryptogamic crusts. Vegetation density is much higher than at the previous unit. This means that much more water is absorbed by the *Stipa* tussocks, which considerably reduces overland flow and improves soil moisture conditions. Water also infiltrates into cracks and joints of the exposed limestone rocks.

6.2 Soil erodibility and water retention within the response units

Quantitative studies of soil erosion on the response units described above have been completed or are in progress. For purposes of comparison and to provide quantitative data for comparing the indicators discussed below, information on the soil aggregation and infiltration behaviour of representative surface soils is shown in Figs. 3 and 4. In Fig. 3 the infiltration characteristics of the different units are illustrated by relationships showing time to ponding at different rainfall intensities. The rapid ponding at low intensities on abandoned land and the difference with a tilled soil illustrates the potential for runoff generation on these surfaces.

6.3 Environmental health indicators for the response units

A simplified list of physical and biological indicators of ecosystem process and function is shown in Tables 2 and 3. Following the procedure in NRC (1994), distinctions were made between areas that are in one of three classes: good (= healthy), intermediate (= at risk) or poor (= unhealthy).

The thirteen indicators were assessed for each response unit. The number of indicators classified as good, poor or intermediate for each landscape response unit, is shown in Table 4.

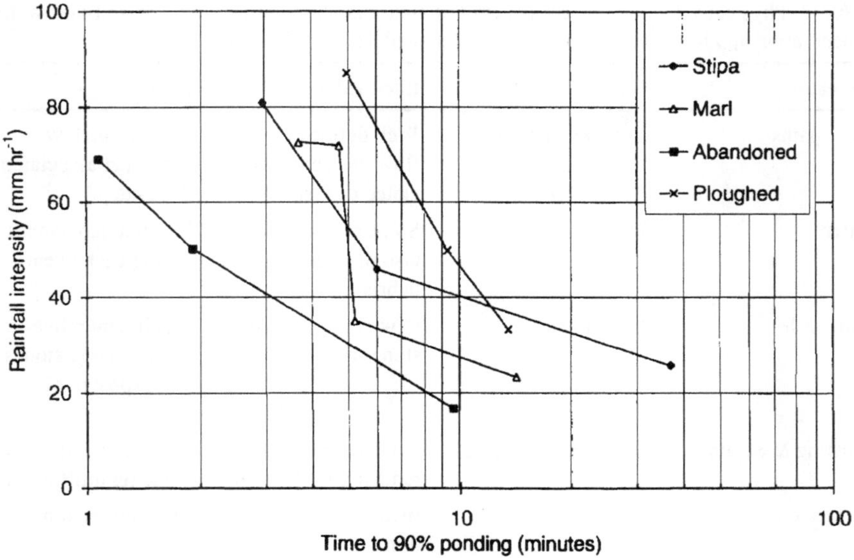

Figure 3. Infiltration characteristics for different land cover units at the Cañada Cazorla area, indicated by the relation between time to 90% water ponded surface and rainfall intensity. (Cammeraat and Imeson, unpublished material).

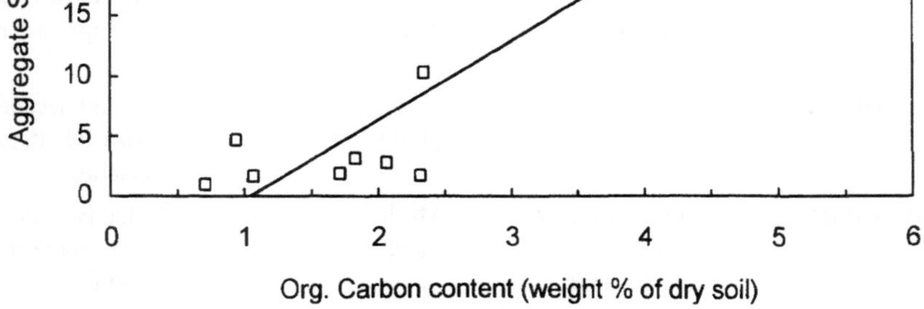

Figure 4. The Relationship between organic carbon and soil aggregate stability for different land units at the Cañada de Cazorla (Cammeraat and Imeson, unpublished material). The aggregate stability index indicates the amount of water drops needed to disrupt 50% of the soil aggregates (see also Cammeraat and Imeson, 1998).

Table 2. Physical indicators of ecosystem function and structure relevant for soil and water conservation applied to the response units. (Adapted from NRC, 1994).

Indicator	Good	Intermediate	Poor
Flow paths	Little evidence of water movement from unit	Well defined small flow lines and associated deposits	Numerous flow paths and associated deposits
Litter	In place	Some displacement also of larger organic debris	Extreme movement during each event
Rainwash	No evidence	Few micro-terraces, stones moved	Significant movement of large stones and exposure of roots
Crusting & Sealing	None or very limited	Crusting obvious, reducing infiltration	Hard crusts strongly reducing infiltration
Surface cover	Soil protected	Incomplete protection	Little protection
Rills	Uncommon	Occasionally present	Very common
Gullies	None	Few but not very active	Numerous active on 20% of length

Table 3. Biotic Indicators of ecosystem function and structure relevant for soil and water conservation applied to the response units. (Adapted from NRC, 1994).

Indicator	Good	Intermediate	Poor
Diversity	Good representation of lifeforms and number of species	1 or 2 lifeforms; Only 30% expected species	One lifeform class dominant; 50% expected species
Structure	Good diversity of height size and distribution of plants and roots	Moderate diversity of plants and root sizes height and distribution	One or two lifeform classes dominant with poor distribution of species and roots
Plant status	Most plants productive and alive	Signs of mortality, production declining	Dead and decadent plants evident, poor production
Nutrient cycle	Mechanisms adequate for plant maintenance	Mechanisms marginally adequate	Mechanisms inadequate to support lifeforms
Grazing pressure	Very little	Obvious but recovery possible	Serious damage, plants unable to recover
Soil fauna	Many ant nests and other soil fauna	Less than 5 nests per 25 m^2, no other soil fauna	Less than 1 ant nest per m^2, no other fauna observed

The region has been suffering a severe drought for the last four years, with average annual rainfalls of <200 mm yr^{-1}. For this reason the biological indicators give low values. On the other hand, the indices 'lack of grazing pressure' and 'soil meso-faunal activity' were usually positive. Grazing was not as intensive as in many other Mediterranean areas and the soil biological activity was usually above class of poor. Furthermore, both the forest and the mattoral are regions dominated by only one life-form and only a few cohorts are usually found. Although these conditions are a fact, a more thorough comparison with conditions on reference sites must be made. It is not surprising, in an area where precipitation is the limiting factor, that the areas that are classified as ecologically healthy are those that receive additions of water from other slope units.

The physical indicators show a trend similar to that of the biological indicators, but the results are sometimes wrong because the indicators are not pertinent to the processes and conditions at the site. For example the use of gully erosion as an indicator is not always valid because conditions do not allow for their formation.

In view of this it seems better not to apply indicators globally, but to only use indicators relevant to the conditions at the response site.

Table 4. Distribution of indicators (out of 7 physical and 6 biological) with good (G), poor (P) or intermediate (I) status for the different response units at Alquería. Response unit abbreviations: cult. = cultivated; ab.= abandoned and acc.= accumulating.

Response unit		Biological indicators			Physical indicators			Combined		
		G	I	P	G	I	P	G	I	P
1A	*Colluvium* cult.	1	1	4	1	0	6	2	1	10
1B	*Colluvium* ab.	0	4	2	1	5	1	1	9	3
1C	*Colluvium* acc.	4	1	1	3	2	2	7	3	3
2	*Stipa* mattoral (1)	3	2	1	2	3	2	5	5	3
3	Shrubs limestone	4	2	0	5	2	0	9	4	0
4	Limestone slope	0	6	0	2	4	1	2	10	1
5	Pine forest	3	3	0	6	1	0	9	4	0
6	Degraded forest	0	2	4	0	1	6	1	5	7
7	Degraded mattoral	0	3	3	1	2	4	1	5	7
8	*Stipa* mattoral (2)	0	5	1	0	6	1	0	11	2

7. CONCLUSIONS

The concepts of desertification and indicators of soil health or quality have much in common. The general indicators that have been developed for these purposes can be adapted to desertification and they are of the same nature. However, the desertification response unit approach highlights that ideally each response unit should have its own specific indicators because the processes that are operating at the patch scale within the response units have a dynamic that is partially determined by constraints imposed by the position of the unit and by its relationship to other units. The same indicator may have a different meaning and a different critical value in different units.

Although this may be seen as a hopeless problem in the search for simple universal indicators, the opposite is the case. By dividing a landscape up into process response units, a rational stratification is obtained for applying indicators according to the processes they reflect. Critical values will depend on the local conditions and should be established from the health of the patterns and structures in the landscape at different scales. Furthermore, it is argued that for the semi-natural areas in particular, the spatial patterns and density of vegetation reflect the long-term interaction with active processes on the hillslope. This makes a translation possible from fine scale field processes and characteristics specific for the response unit scale (hillslope or micro-catchment scale) to far broader scales. The method can therefore help to form a bridge between field research and modeling which needs area-specific data to set initial conditions.

ACKNOWLEDGEMENTS

We gratefully acknowledge the European Commission for financial and scientific support under the Environment and Climate and Environment Programmes (EV5V-CT92-0128 and ENV4 CT95 0115) as part of MEDALUS II and MEDALUS III and the Netherlands Geo-Science Foundation for their support in the Hierarchy in Mediterranean Land Degradation programme. We acknowledge all of those who have discussed this work with us.

REFERENCES

Boardman, J. and Favis-Mortlock, D. 1998. Modelling Soil erosion by water NATO ASI Series 1. Global Environmental Change 55, 531.

Brandt C.J. and Thornes, J.B. 1996. Mediterranean Desertification and Land Use, Wiley, Chichester, 554 pp.

Cammeraat, L.H. and Imeson, A.C. 1998. Deriving indicators of soil degradation from soil aggregation studies in SE Spain and S France. Geomorphology 23, 307–321.

Cammeraat, L.H. and Imeson, A.C. 1999a. A field experiment for up-scaling, hydrological processes under semi-arid conditions: basic outline and preliminary results. Catena. (Submitted).

Cammeraat, L.H. and Imeson, A.C. 1999b. The significance and evolution of soil-vegetation patterns following land abandonment and fire in Spain. Catena. (In press).

Cerdà, A., García-Alvarez, A., Cammeraat, L.H. and Imeson, A.C. 1994. Agregación del suelo en una catena afectada por el abandono del cultivo en la cuenca del Guadalentín (Murcia). I. Estabilidad y distribución de los agregados del suelo. In: Efectos geomorfológicos del abandono de tierras. Eds. J.H. García-Ruiz and T. Lasanta. pp 9–19. Sociedad Española de Geomorfología, Zaragoza.

Cerdà, A., García-Alvarez, A., Cammeraat, L.H. and Imeson, A.C. 1995. Fluctuation estacional y dynamica microbiana en una catena afectada por el abandono del cultivo en la cuenca del Guadalentin (Murcia). Pireneos 145/146, 3–11.

Cerdà, A. 1997. The effect of patchy distribution of Stipa tenacissima L. on runoff and erosion. Journal of Arid Environments 36, 37–51.

Doran, J.W. and Jones, A.J. 1996. Methods for Assessing Soil Quality. SSSA Special Publication 49. SSSA, Madison, 410 pp.

Doran, J.W. and Parkin, T.B. 1996. Quantitative indicators of soil quality: A minimum data set. In: Methods for Assessing Soil Quality. Eds. J.W. Doran and A.J. Jones. pp. 25–38. SSSA, Madison.

FAO 1989. FAO/Unesco Soil Map of the World, Revised Legend. World Resources Report 60, FAO, Rome. Reprinted as Technical Paper 20, ISRIC, Wageningen, 138 pp.

Harris R.F, Karlen, D.L and Mulla, D.J. 1996. A conceptual framework for assessment and management of soil quality and health. In: Methods for Assessing Soil Quality. Eds. J.W. Doran and A.J. Jones. pp. 61–82. SSSA, Madison.

Imeson A.C., Cammeraat, L.H. and Perez-Trejo, F. 1995. Desertification response units. In: Desertification in a European Context Physical and Socio-economic Aspects. Eds. R. Fantechi, R.D. Peter, P. Balabanis and J.L. Rubio. pp. 263–277. Office for Official Publications of the European Community, Luxembourg.

Imeson, A.C., Perez-Trejo, F. and Cammeraat, L.H. 1996. The response of landscape units to desertification. In: Mediterranean Desertification and Land Use. Eds. J. Brandt and J.B. Thornes. pp. 447–469. Wiley, Chichester.

Imeson, A.C., Cammeraat, L.H. and Prinsen, H. 1998, A conceptual approach for evaluating the storage and release of contaminants derived from process based land degradation studies: an example from the Guadalentin basin, SE Spain. Agriculture, Ecosystems and Environment 1265, 1–16.

Imeson, A.C. and Lavee, H. 1998. Investigating the impact of climate change on geomorphological processes: the transect approach and the influence of scale. Geomorphology 23, 219–227.

Karlen, D.L, Mausbach, M.J., Doran, J.W., Cline, R.G., Harris, R.F. and Schuman, G.E. 1997. Soil quality: A concept, definition and framework for analysis. Soil Sci. Soc. Am. J. 61, 4–10.

Kirkby, M.J., Imeson, A.C., Bergkamp, G. and Cammeraat, L.H. 1996. Scaling up processes and models from the field plot to the watershed and regional areas. Journal of Soil and Water Conservation 51(5), 391–396.

Landres, P.B. 1992. Ecological indicators: panacea or liability. In: Ecological Indicators. Eds. D.H. McKenzie, D.H. Hyatt and V.J. McDonald. pp. 1295–1318. Elsevier Applied Science, London & New York.

Lavee, H., Imeson, A.C. and Parientes, S. 1998. The impact of climate change on geomorphology and desertification along a Mediterranean-Arid transect. Land Degradation and Rehabilitation 9, 407–422.

Navarro-Hervas, F. 1991. El sistema hidrografico del GuadalentRn. Consejeria de Politica Territorial, Obras Publicas y Medio Ambiente, Murcia, 256 pp.

NRC 1994. Rangeland Health: New Methods to Classify Inventory and Monitor Rangelands. National Research Council, National Academy Press, Washington, DC.

O'Neill, R.V., DeAngelis, D.L., Waide, J.B. and Allen, T.F.H. 1986. A Hierarchical Concept of Ecosystems. Princeton University Press, New Jersey, 253 pp.

Puigdefabregas, J. and Sanchez, G. 1996. Geomorphological implications of vegetation patchiness on semi-arid slopes. In: Advances in Hillslope Processes. Eds. M.G. Anderson and S.M. Brooks. pp.1027–1060. John Wiley & Sons, London.

Wolman, M.G. and Miller, J.P. 1960. Magnitude and frequency of forces of geomorphic processes. Journal of Geology 68, 54–74.

Agricultural and ecological perspectives of vegetation dynamics and desertification

M. Timm Hoffman
National Botanical Institute, Private Bag X7, Claremont, 7735, Republic of South Africa
Tel: 27 21 762 1166; Fax: 27 21 797 6903; E-mail: <hoffman@nbict.nbi.ac.za>

ABSTRACT Two views of vegetation dynamics and desertification are assessed in terms of their theoretical principles, geographic applicability, perceptions, measurement of degradation and management application. The first view embodies an agricultural perspective developed from a lengthy research programme in the relatively unpredictable Nama-karoo biome of South Africa. The guiding theoretical principle within agricultural research borrows heavily from Clementsian succession theory. Degradation measurement techniques and perceptions of vegetation change have been developed with predictable replacement sequences as the main theme. Relatively little emphasis has been placed on developing a detailed mechanistic understanding of degradation processes in the Nama-karoo. Despite this, much success has been achieved with the approach in developing management programmes aimed at halting or reversing the further degradation of the Nama-karoo. In contrast, the more recent ecological perspective of vegetation dynamics and desertification, emphasizes spatial and temporal unpredictability as its guiding principle. Ironically most ecological research has been carried out in the relatively more predictable Succulent Karoo biome in the south and west regions of South Africa. Small and large-scale disturbances, whether climatic or biological are viewed as key factors determining community composition through time. Detailed demographic models have been produced which emphasize competitive interactions between different guilds of species as they relate to recruitment and survival strategies. Because of the relative isolation of ecological science in general, this research has had relatively little influence on managers in the region.

Key words: desertification, Karoo, land degradation, stability, succession, vegetation change.

1. INTRODUCTION

Until recently, South African perceptions of desertification have been synonymous with those of Karoo rangeland degradation. Soil degradation, bush encroachment and other forms of land degradation (e.g., alien plant invasions, loss of species and plant cover), generally have not formed part of the historical debate on desertification in South Africa (Hoffman, 1997). Instead, it has been the notion of a less productive, semi-arid karroid shrubland expanding its range at the expense of more productive grassland vegetation, that has fuelled much of the alarm over desertification in South Africa (Hoffman and Cowling, 1990). This idea has also had a profound influence on agricultural and environmental policy over the last 100 years (Du Toit et al., 1991; Hoffman et al., 1995). For the Karoo, two main perspectives on vegetation dynamics and rangeland degradation may be recognized. The first refers to an agricultural perspective and the second to an ecological perspective.

Agricultural research has enjoyed a relatively long history in the Karoo. Since 1934 the Department of Agriculture has maintained a research programme at Middleburg (Fig. 1). This programme has tried to solve a range of applied agricultural problems concerned with pasture and rangeland management and animal husbandry practices. Numerous long-term experiments (e.g., O'Connor and Roux, 1995; Roux, 1966) have provided important insights into the vegetation dynamics and land management of the region. Researchers have mostly been drawn from the civil service within the state Department of Agriculture.

Sustained ecological research programmes in the Karoo have, however, developed largely over the last two decades (Cowling, 1986; Werger, 1978). The rise of a national research programme, the Karoo Biome Project (KBP) in 1985, did much to focus ecological research in the region. Although the KBP was short-lived, it nurtured the development of a small ecological research community, with an interest in and dedication to Karoo environmental issues. This interest continues to the present, maintained largely by several university and research institute employees.

From the KBP's inception, attempts were made to integrate agricultural and ecological perspectives and the two research communities have borrowed extensively from each other's work over the last decade (e.g., Milton and Dean, 1996; Milton and Hoffman, 1994). Regular joint research meetings have provided a forum for the exchange of ideas. In one sense, then, the dichotomy between agricultural and ecological perspectives is false. However, I will show that, despite the close working relationship enjoyed by agricultural and ecological researchers, different perspectives on vegetation dynamics and rangeland degradation still exist. This is partly because agricultural research has focussed mostly on the dynamics, management and degradation of areas within the eastern parts of the Nama-karoo biome, while ecological research has been more centred on processes within the Succulent Karoo biome. Different research objectives organized around different theoretical principles have also contributed to the apparent dichotomy. By emphasizing these differences I will try to provide a clearer understanding of Karoo vegetation dynamics and its degradation as a whole.

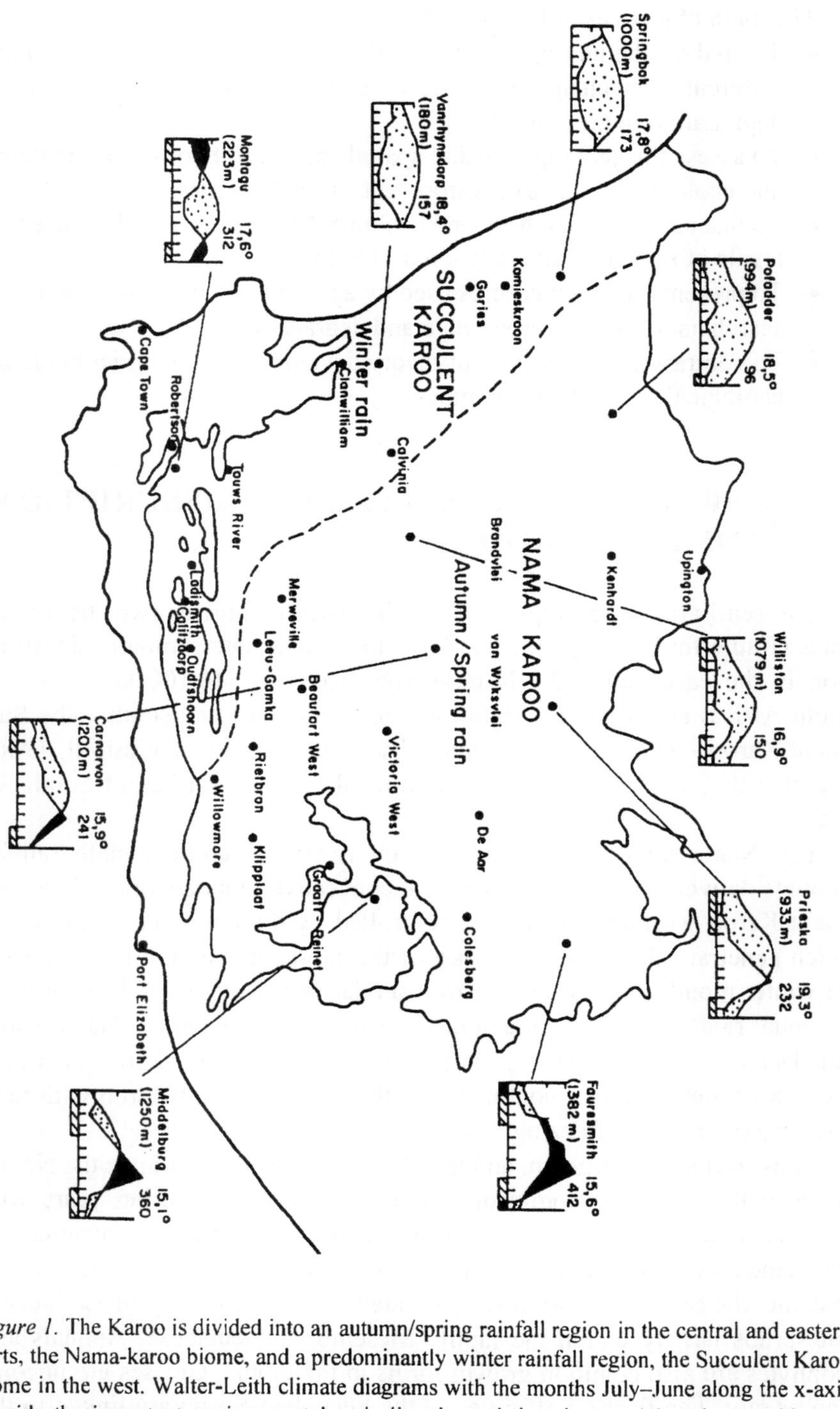

Figure 1. The Karoo is divided into an autumn/spring rainfall region in the central and eastern parts, the Nama-karoo biome, and a predominantly winter rainfall region, the Succulent Karoo biome in the west. Walter-Leith climate diagrams with the months July–June along the x-axis provide the important environmental and climatic statistics characterizing the two biomes. (From Hoffman and Cowling, 1987).

The main objectives of this contribution are:

- To analyze the important theoretical principles which have guided the different research approaches to vegetation dynamics and rangeland degradation in the Karoo;
- To assess the geographical and spatial applicability of key agricultural and ecological models of Karoo vegetation dynamics;
- To discuss the agricultural and ecological perceptions of the current levels of rangeland degradation in the Karoo;
- To present the key measures used by agricultural and ecological researchers to assess Karoo rangeland degradation;
- To contrast the application of theory developed within agricultural and ecological research programmes.

2. SETTING THE SCENE: A BRIEF DESCRIPTION OF THE KAROO

The semi-arid Karoo region of South Africa comprises two distinct biomes (Rutherford, 1997) (Fig. 1). A predominantly late summer rainfall region in the east, called the Nama-karoo, occupies 346,100 km^2 (28% of South Africa) and a winter rainfall region in the west and south – the Succulent Karoo – occupies 81,100 km^2 (7%) (Rutherford and Westfall, 1986). Together they comprise more than a third of the land surface of South Africa.

The Nama-karoo is characterized by relatively unpredictable rainfall amounts delivered mostly via localized, convective thunderstorms. The Succulent Karoo weather systems are controlled by the circumpolar westerlies which penetrate along the west coast of the sub-continent and inland during the winter months depositing rain over a relatively wide front. The variation in annual rainfall for a given amount is generally higher in the Nama-karoo and decreases with increasing rainfall (Fig. 2). The Nama-karoo also possesses a greater degree of continentality than the Succulent Karoo with temperatures often below freezing in winter.

Non-succulent, evergreen, and leaf deciduous shrubs dominate the Nama-karoo, with grasses co-dominant in the east, especially during years with good summer rains. It has strong floristic affinities with the Savanna and Grassland biomes to the north and east (Rutherford, 1997). As suggested by its name, the Succulent Karoo is dominated by a wide variety of leaf succulent shrubs largely within the family Mesembryanthemaceae. Annuals and geophytes are also common growth forms in the biome. Grasses are uncommon or rare. The floristic affinities of the Succulent Karoo are linked to the Fynbos biome to the south. The two Karoo biomes share only 28% of their species (Rutherford, 1997). More detailed descriptive accounts of the two biomes may be found in Milton et al. (1997) and Palmer and Hoffman (1997).

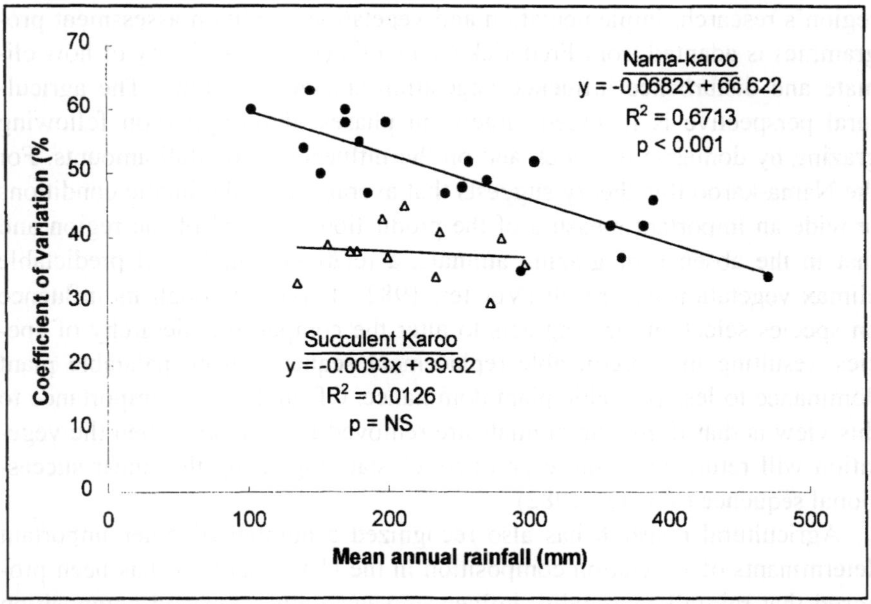

Figure 2. The relationship between mean annual rainfall (MAR) (mm) and coefficient of variation (%) in MAR for 12 stations in the Succulent Karoo biome and 20 stations in the Nama-karoo biome. (After Hoffman and Cowling, 1987).

3. IMPORTANT THEORETICAL PRINCIPLES

Agricultural research has been carried out largely in the eastern and central parts of the Nama-karoo biome. One of its main tasks over the years has been to develop key principles for sustainable land use practices within the region. Farmers and policy makers within national government are important clients and research has had to be generalized so that management systems for large areas can be applied. Amongst many other priorities, the focus in agricultural research on Karoo rangelands has been on: (1) Identifying relatively homogenous farming areas as key planning tools (Vorster et al., 1987); (2) Developing carrying capacity estimates which apply over decades and which are linked to mean annual rainfall (Van den Berg, 1983); (3) Developing veld condition assessment tools which rely on generalized estimates of "excellent", "fair", "poor" and "very poor" vegetation condition (Vorster, 1982).

Although this is the most variable of the two biomes in terms of annual rainfall (Fig. 2), the most important theoretical principle underpinning the region's research, implementation and vegetation condition assessment programmes is adapted from Frederick Clement's ecological theory of how climate and disturbance influence vegetation change over time. The agricultural perspective is focussed largely on phases of retrogression following grazing by domestic livestock and on the influence of rainfall amounts. For the Nama-karoo this theory suggests that average annual climatic conditions provide an important measure of the production potential of the region and that in the absence of grazing animals, a relatively stable and predictable climax vegetation will result (Vorster, 1982). Largely through its influence on species selection, grazing acts to alter the competitive hierarchy of species, resulting in a predictable replacement sequence from palatable plant dominance to less palatable plant dominance. Of fundamental importance to this view is that if grazing animals are removed from an area then the vegetation will return to a "more productive" state higher up the linear successional sequence (Vorster, 1982).

Agricultural research has also recognized a number of other important determinants of vegetation composition in the Nama-karoo. It has been proposed that rainfall seasonality influences vegetation cover and composition profoundly, largely through its effect on grass and shrub cover (Roux, 1966). High summer rainfall amounts in the eastern Karoo are correlated with high annual totals, and both are positively related to an increase in grass cover (Hoffman et al., 1990; O'Connor and Roux, 1995; Roux, 1966). Similarly, Roux (1966, 1968) has suggested that autumn rain favours shrub cover. Because rainfall seasonality influences shrub and grass cover, the competitive interactions between grasses and shrubs (Hoffman et al., 1990) are also worth investigating further, despite O'Connor and Roux's (1995) rejection of the hypothesis that these two growth forms compete for space and resources.

The influence of rainfall seasonality and the timing of grazing on grass and shrub cover provides one of the key principles for the management system proposed by the Department of Agriculture (Hoffman, 1988a; Roux, 1968). By grazing paddocks at different times of the year it has been shown that the proportion of shrubs and grasses in the vegetation can be determined in a predictable way. Summer grazing favours an increase in shrub cover, while winter grazing promotes grass cover. The Group Camp System (Roux, 1968), a deferred grazing approach to Nama-karoo rangeland management, is based on these principles.

Milton and Hoffman (1994) have developed a state-and-transition model of the dynamics of Nama-karoo vegetation. It incorporates both grazing and climatic effects although details of the transitions are stated as hypotheses at this stage. However, despite the success of its management recommendations, the agricultural perspective has not as yet provided a comprehensive mechanistic understanding of the grazing and climatic influences on vegetation change in the Nama-karoo.

With notable exceptions (e.g., Bond et al., 1994), most ecological research in the Karoo has been carried out within the confines of the Succulent Karoo biome. It is ironic that the biome which is more predictable in terms of rainfall variation, should generate research perspectives which emphasize the importance of rainfall variation, small- and large-scale disturbances and patchiness as key theoretical principles for understanding the vegetation of the region (Milton et al., 1997). Whereas agricultural research has always had pressing management problems to solve, ecological research has enjoyed far fewer constraints and pressures in the development of its theory. Both have borrowed extensively from international perspectives, but ecological research has drawn from a wider range of disciplines (e.g., archaeology, environmental history, climatology, zoology (especially ornithology), botany) operating across greater temporal and spatial scales than has been common within agricultural research. These differences are reflected in their respective views on vegetation dynamics and rangeland degradation.

Karoo ecologists have generally not emphasized the stability of Karoo ecosystems and the concept of a stable "climax" vegetation, as determined by average annual rainfall conditions. On the contrary, they have been critical of this view as an effective organizing principle for the region (Hoffman and Cowling, 1987). Their rejection of parts of the theory is supported by a number of recent studies which show that the removal of animals from a paddock does not result in a return to the hypothesized climax vegetation, even after many years (Fairall and Le Roux, 1991; Milton, 1992; Stokes, 1994). Ecologists have tried to discover why this is the case and have adopted a far stronger demographic approach to understanding the vegetation dynamics of the Karoo than is evident in the agricultural literature. The influence of climate and grazing on the mortality, phenology (especially flowering and seed production), recruitment and survival of individuals has received much attention (Esler, 1993; Milton et al., 1997). A more detailed and mechanistic state-and-transition model is thus available for the Succulent Karoo than for the Nama-karoo biome (Milton and Hoffman, 1994).

While Noy-Meir's (1973) notion of arid systems being "event-driven" received considerable attention in the early years of the Karoo Biome Project (Hoffman, 1988b), the key processes structuring Succulent Karoo biome ecosystems have emerged in a series of models for the region. Yeaton and

Esler (1990), for example, produced a cyclical succession model in which predictable replacement sequences of succulent and non-succulent shrubs were identified (see also Milton et al., 1997). Competitive hierarchies are inferred from nearest-neighbour analyses, and the role of nutrient enriched soil mounds, germination microsites and small-scale, animal disturbances are central to their model.

The most comprehensive syntheses of Karoo vegetation dynamics and degradation emphasize the role of rainfall stochasticity and its effect on key population processes (Wiegand and Milton, 1996; Wiegand et al., 1995). The life histories of five dominant species and long-term monthly rainfall data were used to simulate the dynamics of a shrub ecosystem in the Succulent Karoo. The model shows that the spatial and temporal small-scale patch dynamics of this ecosystem are strongly influenced by rainfall conditions. Long, quasi-stable conditions lasting decades, alternate with episodic and discontinuous changes in community composition. This behaviour is determined largely by the specific responses of individual species to a particular sequence of rainfall events and by the density of plants at the time (Wiegand et al., 1995). Gaps, created both by drought and small fossorial mammal disturbances, appear to be crucial determinants of change in the ecosystem. Simulated resting of heavily grazed vegetation did not result in any significant changes in species composition over 60 years. Intervention techniques such as shrub removal also proved ineffective in restoring the vegetation to its full productive potential (Wiegand and Milton, 1996).

The relative influence of event-driven versus continuous processes is an important question faced by managers. Some recent work in Australia's drylands suggests that continuous processes may be as important as event-driven or episodic processes and that the former are under some control by management, whereas the latter are much less so (Watson et al., 1996, 1997a, 1997b). Multiple timescales, from years to decades or longer, all need to be considered in adaptive management programmes which consider change as being both continuous and event-driven and which prepare the resource to respond optimally to a given event (Watson et al., 1996).

4. GEOGRAPHICAL AND SPATIAL APPLICABILITY OF THE DYNAMIC MODELS

Agricultural research has developed management programmes that can be applied over large areas. However, the Nama-karoo contains a wide variety of vegetation types with different climates, soils and species composition (Palmer and Hoffman, 1997). The general model for vegetation in the eastern part of the Nama-karoo with a mixture of grass and shrubs has not been assessed for other parts of the biome. It is likely, however, to provide only a limited understanding of the vegetation dynamics of the region's arid grasslands in the west or the montane grasslands which exist at higher elevations in the east of the biome.

Similarly, the models of vegetation dynamics developed for the southern part of the Succulent Karoo biome are likely to have limited value in understanding the dynamics of the vegetation in the west. The dominance of *Pteronia pallens* and other long-lived shrubs does not occur throughout the biome. In the western regions of the Succulent Karoo, leaf succulent shrubs, with life spans of one to two decades only, dominate the landscape. The turnover in individuals and species within a community appears much more rapid although some of the same factors, such as drought and disturbance, are likely to remain as important controls of population processes (Cowling and Hilton-Taylor, in press; von Willert et al., 1990).

5. PERCEPTIONS OF DEGRADATION

Although few supporting data have been published, the general perception within the agricultural research community is that the Karoo has changed from a pre-colonial perennial grassland to its current mix of relatively unpalatable grasses and perennial shrubs in historical times (Table 1) (Roux, 1981; Roux and Theron, 1987; Roux and Vorster, 1983). In nearly all respects, the overlapping or "serial" stages of degradation envisaged for the Karoo over the last century, mirror the veld condition classes used to measure the current status of vegetation (Vorster, 1982) (Fig. 3). Like Acocks (1953), Roux and Theron (1987) discount climatic change as a primary cause of land degradation in the region. They suggest instead that the impact of the extensive small stock farming industry, especially the impact of overstocking, is to blame for the current poor state of the region's vegetation.

Key arguments, which summarize the ecological perceptions of desertification in the Karoo, have recently been presented (Dean et al., 1995; Hoffman 1995, 1997). Overstocking is also regarded as the prime cause of rangeland degradation, but there is disagreement as to the timing of such degradation, as well as to the current state of the vegetation in the Karoo, relative to historical times. A wide range of palynological, archaeological, historical and ecological methodologies have been used to determine the nature of pre-colonial Nama-karoo environments (see Hoffman, 1995). Most disciplines agree that the Nama-karoo was more grassy in the past, but disagree on the timing of the shift from grassy to shrubby conditions. Some suggest it paralleled the rise of the small stock industry in the Nama-karoo while others find little evidence for this (Hoffman, 1997).

Table 1. An agricultural perspective of the main phases of vegetation change in the Karoo, including dates and key processes involved. (After Roux and Vorster, 1983; Roux and Theron, 1987).

Phase	Period	Description and key processes
1. Primary degradation	1880–1900/20	Loss of cover, particularly of perennial climax grasses, short-lived grasses, some palatable dwarf shrubs
2. Denudation	1900/20–1940/50	Further removal of palatable grasses and dwarf shrubs and start of unpalatable dwarf shrub dominance. High rates of erosion. Process exacerbated by severe droughts
3. Re-vegetation	1940/50–1970/80	Increase in cover of particularly unpalatable dwarf shrubs, short-lived grasses and herbaceous annuals. Taller shrubs start to invade
4. Secondary degradation	1970/80–2000	Relatively stable grazing disclimax, dominated by a few undesirable and unpalatable species. Perennial grass absent, herbaceous annuals common
5. Desertification	Beyond 2000	Vegetation entirely destroyed except for one or two hardy shrubs and herbaceous annuals. Soil completely exposed to erosive agents. Exotic species invade

Perhaps the greatest disagreement concerns the current state of the Karoo environment relative to historical times. One recent review (Hoffman and Cowling, 1990) found little evidence to support Acocks' (1953) earlier gloomy predictions of a rapidly expanding karroid shrubland. The "expanding Karoo hypothesis" has dominated the popular, semi-scientific and scientific literature since its first airing but has rarely, until recently, been subjected to detailed scrutiny. Using historical landscape descriptions, matched photographs and vegetation survey records spanning 30 years, Hoffman and Cowling (1990) suggested that the Karoo at the time of European colonization during the 19th century, was already predominantly a shrubland, and that few physiognomic changes were evident in the data. If anything, the region had experienced an "improvement" as to total cover and an increase in grass cover in the latter part of the 20th century. The sustained efforts of the Department of Agriculture via their research, education and especially their farmer support programmes (e.g., drought subsidies), were cited as key reasons why the Karoo maintained a better state of ecosystem health today, than certainly is evident in the photographs taken in the first three decades of this century (Hoffman et al., 1995).

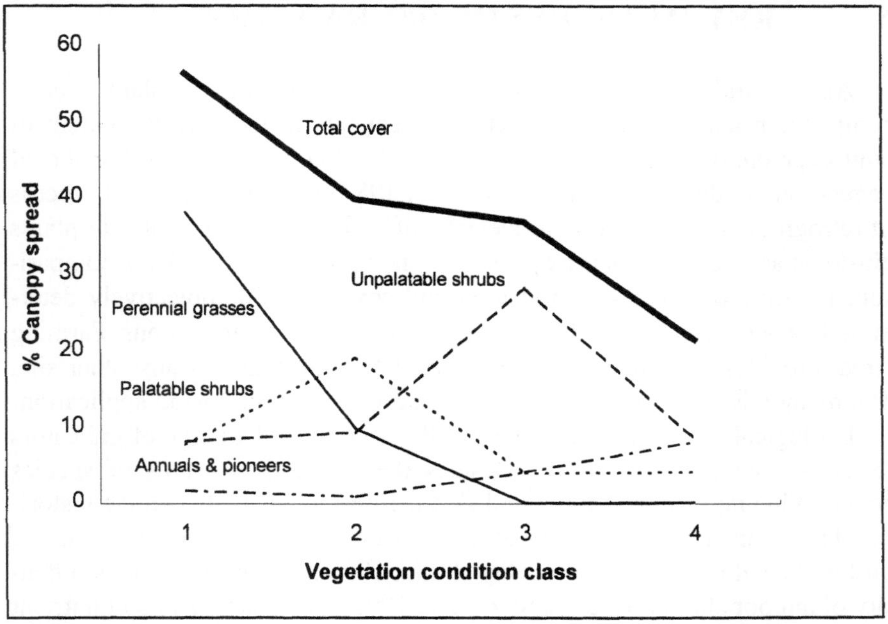

Figure 3. The percentage canopy spread of four growth forms and total vegetative cover within four vegetation classes for vegetation in the eastern part of the Nama-karoo biome. The vegetation condition classes are: 1=Excellent; 2=Fair; 3=Poor; 4=Very poor. (Data from Vorster, 1982).

A comprehensive analysis of the historical stocking rates for magisterial districts in the Karoo has, however, suggested that there has indeed been widespread degradation over the historical period (Dean and Macdonald, 1994). An analysis of stable soil isotopes has also supported the idea of a dramatic decline in C4 grasses in the last several decades (Bond et al., 1994). Brief responses to these arguments have been made (Hoffman et al., 1995). However, it is evident that more dynamic models of desertification should be developed for the eastern Karoo (Dean et al., 1995), particularly since grass and shrub cover is so clearly influenced by seasonal rainfall patterns (Hoffman et al., 1990; Milton and Hoffman, 1994; O'Connor and Roux, 1995; Roux, 1966). During periods of high summer rainfall grass cover is unlikely to decline and degradation models should account for these strong seasonal influences.

6. KEY MEASURES OF DEGRADATION

Agricultural research has emphasized vegetation cover, plant species composition and livestock production parameters to measure the extent of land degradation in an area. For example, the Ecological Index Method of vegetation condition assessment (Vorster, 1982) uses the perceived process of retrogression in the Karoo to identify different ecological groups of plants clustered according to their response to grazing pressure (Fig. 3). A comparison, in terms of species composition and cover, with a subjectively determined benchmark site within the same Relatively Homogenous Farming Area, provides a relative measure of vegetation condition at any other site. The method is relatively easy to use and has enjoyed widespread application.

Ecological research on the other hand has employed a range of indicators to assess vegetation condition. Besides the standard measures of species cover and composition (Milton and Dean, 1996), other tools include historical changes in stocking rates (Dean and Macdonald, 1994), the abundance of bird nests and nest sites at a locality (Dean, 1996), the composition and density of ant populations (Dean and Yeaton, 1993), and changes in soil nutrient content and mycorrhizal abundance (Allsopp, 1997). Because they require fairly specialized equipment and knowledge, they have rarely been used over wide areas and not for any length of time (for proposed alternatives see Tongway and Hindley, this volume; Imeson and Cammeraat, this volume).

7. APPLICATION OF THEORY

Agricultural research has, of necessity, adopted an applied approach to their research programme. Research findings inform decision-makers and politicians within the Department of Agriculture and form part of the general body of knowledge taught at agricultural colleges. It is also transferred to farmers via the media and an active agricultural extension service. This information influences national resource conservation agencies and ultimately feeds into the law-making process, which is charged with the wise use of the country's resources. Agricultural research, therefore, operates within a large supportive framework and has been successfully applied over the last half century.

Despite its contribution to our understanding of Karoo ecosystem structure and function over the last 20 years, ecological research has seldom fed directly into management or policy-related structures. Scientific knowledge has largely been created for consumption within the scientific community.

Table 2. Summary of main agricultural and ecological perceptions of vegetation dynamics and rangeland degradation in the semi-arid Karoo, South Africa.

	Agricultural perspective	Ecological perspective
Research history	Relatively lengthy (since 1934) and carried out largely by rangeland scientists from the state Department of Agriculture. Focus has been on the eastern part of the Nama-karoo biome	Relatively short (last one to two decades) and carried out largely by biologists from academic institutions. Focus has mostly been on the Succulent Karoo biome
Important theoretical principles	• Succession theory dominant paradigm • Climatic equilibrium and uniformity emphasized, although rainfall seasonality has important influence • Role of grazing animals crucial • Limited mechanistic understanding of degradation	• More accepting of event-driven paradigm of arid ecosystem function • Variation in climate and patchiness emphasized • Role of grazing animals crucial but other organisms also included as being important influences on ecosystem degradation • Demographic and population processes provide mechanistic understanding
Geographical and spatial applicability of dynamic models	Relatively good for the eastern parts of the Nama-karoo biome but application in other parts of the Karoo is not known (e.g., arid or montane grasslands)	Poorly tested outside of the *Pteronia pallens* dominated shrublands of the southern Karoo; indications are that it may have local applicability only
Are Karoo rangelands degraded?	Generally perceived to be degraded, although supporting data are scarce	Extensive multi-disciplinary debate. Less idealised viewpoint about historic condition and less optimistic about potential for recovery
Key measures of degradation	Vegetation cover, plant species composition and livestock production	Wide range of measures including vegetation cover, plant species composition, stock numbers, biodiversity and impacts on soil processes
Application of theory	Relatively good as it is well linked to state instruments of policy, agricultural extension and education, resource conservation and the law	Relatively poorly applied, with limited outlet beyond the scientific research community; may reflect relatively short history

This appears to be changing however, as popular translations of scientific studies (Milton and Dean, 1996) make ecological knowledge more accessible to a wider audience. The forthcoming synthesis of Karoo research (Dean and Milton, in press) should also go a long way to publicize the current state of ecological knowledge of the Karoo. Ecologists are now also contributing to national environmental policy initiatives as a more open South African society attempts to deal with crucial land use and land degradation issues facing it in the coming decades.

8. CONCLUSIONS

The most important differences between the agricultural and ecological perspectives are summarized in Table 2. The somewhat divergent ecological and agricultural perspectives have strengthened our understanding of Karoo ecosystem structure and functioning. The deliberate borrowing from, and critique of each other's most cherished theories has also helped to create an intellectually active community of students and researchers all dedicated to the management and conservation of a biologically fascinating area.

REFERENCES

Acocks, J.P.H. 1953. The veld types of South Africa. Mem. Bot. Surv. S. Afr. 28, 1–128.

Allsopp, N. 1997. Soil change associated with small stock production and cultivation in the Paulshoek communal area, Leliefontein, Namaqualand. Unpublished Final Project Report. Range and Forage Institute, Agricultural Research Council, Pretoria.

Bond, W.J., Stock, W.D. and Hoffman, M.T. 1994. Has the Karoo spread? A test for desertification using carbon isotopes from soils. S. Afr. J. Sci. 90, 391–397.

Cowling, R.M. 1986. A description of the Karoo Biome Project. Pretoria, CSIR, S. Afr. Nat. Sci. Prog. Rep. No. 122, 1–43.

Cowling, R.M. and Hilton-Taylor, C. (In press). Plant biogeography, endemism and diversity. In: The Karoo: Ecological Patterns and Processes. Eds. W.R.J. Dean and S.J. Milton. Cambridge University Press, Cambridge.

Dean, W.R.J. 1996. Where birds are rare or fill the air: The Protection of the Endemic and the Nomadic Avifaunas of the Karoo. Ph.D. thesis. University of Cape Town, Cape Town.

Dean, W.R.J. and Macdonald, I.A.W. 1994. Historical changes in stocking rates of domestic livestock as a measure of semi-arid and arid rangeland degradation in the Cape Province, South Africa. J. Arid Environ. 26, 281–298.

Dean, W.R.J. and Milton, S.J. (In press). Pattern and Process in the Karoo, South Africa. Cambridge Univ. Press, Cambridge.

Dean, W.R.J. and Yeaton, R.I. 1993. The influence of harvester and Messor capensis nest mounds on the productivity and distribution of some plant species in the southern Karoo, South Africa. Vegetatio 106, 21–35.

Dean, W.R.J., Hoffman, M.T., Meadows, M.E., and Milton, S.J. 1995. Desertification in the semi-arid Karoo, South Africa: review and reassessment. J. Arid Environ. 30, 247–264.

Du Toit, P.F., Aucamp, A.J. and Bruwer, J.J. 1991. The national grazing strategy of the Republic of South Africa. Objectives, achievements and future challenges. J. Grassl. Soc. Sth Afr. 8, 126–130.

Esler, K.J. 1993. Vegetation Patterns and Plant Reproductive Processes. Ph.D. Thesis, Univ. of Cape Town, Cape Town.

Fairall, N. and Le Roux, A. 1991. Game management in arid areas: the non-equilibrium alternative. In: Wildlife Production: Conservation and Sustainable Development. Eds. L.A. Renecker and R.J. Hudson. University of Alaska, Fairbanks.

Hoffman, M.T. 1988a. Rationale for karoo grazing systems: criticisms and research implications. S. Afr. J. Sci. 84, 556–559.

Hoffman, M.T. 1988b. The fourth annual research meeting of the Karoo Biome Project: In search of a paradigm. S. Afr. Inst. Ecol. Bull. 7(2), 23–29.

Hoffman, M.T. 1995. Environmental history and the desertification of the Karoo, South Africa. Gionarle Botanico Italiano 129(1), 261–273.

Hoffman, M.T. 1997. Human impacts on vegetation. In: Vegetation of Southern Africa. Eds. R.M. Cowling, D.M. Richardson and S.M. Pierce. pp. 507–534. Cambridge University Press, Cambridge.

Hoffman, M.T. and Cowling, R.M. 1987. The physiognomy, phenology and demography of Karoo plants. In: Karoo Ecology: a Preliminary Synthesis. Part II. Eds. R.M. Cowling and P.W. Roux. S. Afr. Nat. Sci. Prog. Rep. No. 142, pp. 1–34.

Hoffman, M.T. and Cowling, R.M. 1990. Vegetation change in the semi-arid, eastern Karoo over the last two hundred years: An expanding Karoo – fact or fiction? S. Afr. J. Sci. 86, 286–294.

Hoffman, M.T., Barr, G.D. and Cowling, R.M. 1990. Vegetation dynamics in the semi-arid eastern Karoo, South Africa: the effect of seasonal rainfall and competition on grass and shrub basal cover. S. Afr. J. Sci. 86, 462–463.

Hoffman, M.T., Bond, W.J. and Stock, W.D. 1995. Desertification of the eastern Karoo, South Africa: conflicting palaeoecological, historical and soil isotopic evidence. Environmental Monitoring and Assessment 37, 1–19.

Milton, S.J. 1992. Effects of rainfall, competition and grazing on flowering of *Osteospermum sinuatum* (Asteraceae) in arid Karoo shrubland. J. Grassl. Soc. Sth Afr. 9, 158–164.

Milton, S.J. and Dean, W.R.J. 1996. Karoo Veld: Ecology and Management. Agricultural Research Council, Pretoria.

Milton, S.J. and Hoffman, M.T. 1994. The application of state-and-transition models to rangeland research and management in arid succulent and semi-arid grassy Karoo, South Africa. Afr. J. Range For. Sci. 11, 18–26.

Milton, S.J., Yeaton, R.I, Dean, W.R.J. and Vlok, J.H.J. 1997. Succulent Karoo. In: Vegetation of Southern Africa. Eds. R.M. Cowling, D.M. Richardson and S.M. Pierce. pp. 131–166. Cambridge University Press, Cambridge.

Noy-Meir, I. 1973. Desert ecosystems: environment and producers. Ann. Rev. Ecol. Sys. 4, 25–51.

O'Connor, T.G. and Roux, P.W. 1995. Vegetation changes (1949–71) in a semi-arid, grassy dwarf shrubland in the Karoo, South Africa: influence of rainfall variability and grazing by sheep. J. Appl. Ecol. 32, 612–626.

Palmer, A.R. and Hoffman, M.T. 1997. Nama-karoo. In: Vegetation of Southern Africa. Eds. R.M. Cowling, D.M. Richardson and S.M. Pierce. pp. 167–188. Cambridge University Press, Cambridge.

Roux, P.W. 1966. Die uitwerking van seisoenreëval en beweiding op gemengde karooveld. Proc. Grassl. Soc. S. Afr. 1, 103–110.

Roux, P.W. 1968. Principles of veld management in the Karoo and the adjacent dry sweet-grass veld. In: The Small Stock Industry in South Africa. Ed. W.J. Hugo. pp. 318–340. Government Printer, Pretoria.

Roux, P.W. 1981. Vegetation change in the Karoo region. Karoo Agric 1(5), 15–16.

Roux, P.W. and Theron, G.K. 1987. Vegetation change in the Karoo biome. In: Karoo Ecology: a Preliminary Synthesis. Part II. Eds. R.M. Cowling and P.W. Roux. S. Afr. Nat. Sci. Prog. Rep. No. 142, pp. 50–69.

Roux, P.W. and Vorster, M. 1983. Vegetation change in the Karoo. Proc. Grassl. Soc. S. Afr. 18, 25–29.

Rutherford, M.C. 1997. Categorization of the biomes. In: Vegetation of Southern Africa. Eds. R.M. Cowling, D.M. Richardson and S.M. Pierce. pp. 91–98. Cambridge University Press, Cambridge.

Rutherford, M.C. and Westfall, R.H.1986. Biomes of southern Africa – an objective categorization. Mem. Bot. Surv. Sth. Afr. 54, 1–98.

Stokes, C.J. 1994. Degradation and Dynamics of Succulent Karoo Vegetation. M.Sc. thesis, Univ. of Natal, Pietermaritzburg.

Van den Berg, J.A. 1983. The relationship between the long term average rainfall and the grazing capacity of natural veld in the dry areas of South Africa. Proc. Grassl. Soc. S. Afr. 18, 165–167.

Von Willert, D.J., Eller, B.M., Werger, M.J.A. and Brinckmann, E. 1990. Desert succulents and their life strategies. Vegetatio 90, 133–143.

Vorster, M. 1982. The development of the Ecological Index Method for assessing veld condition. Proc. Grassl. Soc. S. Afr. 17, 84–89.

Vorster, M., Becker, H.R. and Greyling, J.S. 1987. Ordination of land types in the Karoo region into reasonably homogenous farming areas based on vegetation and environmental factors. J. Grassl. Soc. Sth Afr. 4, 13–17.

Watson, I.W., Burnside, D.G. and Holm, A.M. 1996. Even-driven or continuous: which is the better model for managers? Rangeland Journal 18, 351–369.

Watson, I.W., Westoby, M. and Holm, A.M. 1997a. Demography of two shrub species from an arid grazed ecosystem in Western Australia 1983–93. Journal of Ecology 85(6), 815–832.

Watson, I.W., Westoby, M. and Holm, A.M. 1997b. Continuous and episodic components of demographic change in arid zone shrubs: models of two Eremophila species from western Australia compared with published data on other species. Journal of Ecology 85(6), 833–846.

Werger, M.J.A. 1978. Biogeography and Ecology of Southern Africa. Junk, The Hague.

Wiegand, T. and Milton, S.J. 1996. Vegetation change in semiarid communities. Vegetatio 125, 169–183.

Wiegand, T., Milton, S.J. and Wissel, C. 1995. A simulation model for a shrub ecosystem in the semiarid Karoo, South Africa. Ecology 76, 2205–2221.

Yeaton, R.I. and Esler, K.J. 1990. The dynamics of a succulent karoo vegetation. Vegetatio 88, 103–113.

The United Nations data bases on desertification

W. Franklin G. Cardy
Executive Coordinator, Natural Resources and Director, Land[1]. *United Nations Environment Programme*
Tel: 1 202 458 8225; Fax: 1 202 473 8249; E-mail: <fcardy@worldbank.org>

ABSTRACT Although debate raged during the 1970s and '80s on how to define desertification, a number of related databases were developed. Most notable was the Global Assessment of Human-Induced Soil Degradation (GLASOD). New assessments have been carried out regionally in Asia and at national level in a number of other countries. Based on these, a new, extensively revised edition of the World Atlas of Desertification was published by UNEP in 1997. Work continues on improving regional and national assessments, and on identification of key land quality indicators. Databases established by United Nations Organisations and others are described in this chapter and referenced. Recommendations for improving our knowledge of the extent and impact of desertification are made.

Key words: databases, desertification, drylands, environment information, land degradation, soils.

1. BACKGROUND

The United Nations Environment Programme (UNEP) and its partners worked for twenty years to get desertification recognized as a significant global issue, to achieve international agreement on its definition, to improve the desertification databases and to bring about a greater effort in implementing preventive measures.

One of the major problems in assembling a good database on desertification during the 1970s and 1980s was the lack of consensus on how to define desertification. Over one hundred definitions have been catalogued (Barraclough, 1995). Without clear agreement on definition, there was no possibility of achieving a meaningful assessment database.

Desertification has now been defined, by international agreement in the United Nations Convention to Combat Desertification (CCD) as "land degradation in arid, semi-arid, and dry sub-humid areas resulting from various factors including climatic variations and human activities" (UN CCD, 1994).

---

1 Since 1998, Senior Water Resource Management Specialist, AFTU2, World Bank, Washington.

These areas cover 40% of the land surface of the globe and are the home of more than one billion people (UNDP/UNSO, 1997). These are "the susceptible drylands" (UNEP, 1992a, 1997).

There is still much to be done to improve the science of this subject. The definition of "desertification" is still debated and there is still a lack of clarity regarding features which distinguish various categories of land degradation. These include physical erosion and degradation, soil degradation (structural and chemical), vegetation degradation, human-induced soil degradation, and climate-induced degradation (UNEP, 1997). In addition, there remain variations in the way the climate zones are defined.

The first world map of desertification was produced by UNESCO, UNEP, FAO and WMO, in time for the United Nations Conference on Desertification (Nairobi, 1977). Most of the subsequent national, regional and global assessments of Desertification undertaken by UNEP and others were based on the FAO/UNEP (1984) "Provisional Methodology for Assessment and Mapping of Desertification". The "provisional methodology" was later made use of by UNEP and its partners in the 1987–1990 period, to produce the first Global Assessment of human-induced Soil Degradation (GLASOD). Various databases relating to desertification were prepared during the 1980s (see Tables for details).

A prerequisite for an acceptable scientific global database is agreement on definition of the subject being studied. In early 1990, UNEP brought together a team of international experts and international organisations to propose an internationally acceptable definition of desertification. Their definition was slightly modified and then adopted at the 'Earth Summit' at Rio in 1992 (Agenda 21, UNCED, 1992). This definition of desertification (presented earlier in this section) is now embodied in the CCD. The definition delimits the areas where desertification occurs and addresses its causes, emphasizing climatic aspects (e.g., short-term droughts, long-term climate fluctuations), and human-induced land degradation, while not ruling out other factors.

The first edition of the World Atlas of Desertification, published in 1992 to coincide with the 'Earth Summit', displayed the existing state of knowledge of desertification and of its extent and possible solutions. It demonstrated that desertification is a major economic, social, and environmental problem affecting more than 110 countries in all regions of the world.

Agenda 21 included a chapter (12) on Managing Fragile Ecosystems: Combating Desertification and Drought. This chapter recommended, inter alia, a better determination of the nature, extent and socio-economic impacts of desertification at local and national levels. It also recommended the establishment of an inter-governmental negotiating committee to prepare a convention to combat desertification.

Table 1. DESIS. The UNEP Desertification Control Information System. The United Nations Environment Programme (UNEP), Desertification Control Programme Activity Centre (DC/PAC) established the Desertification Control Information System (DESIS) in 1987. It is composed of seven databases, available on diskettes from UNEP. (www.unep.org).

Bibliographic databases:
- DESBIB (World Desertification Bibliography Database): A referral database with 3896 references to conventional and non conventional documents on desertification and its control. From 1967 to 1988. Indexed by author, subject and geographical descriptors
- BIWIND: A DESIS database on wind erosion and its control (3665 references)
- KEYS: Desertification Thesaurus Database, 306 key term entries

Projects and activities databases:
- DEPRO: 62 UNEP desertification control projects
- ACWIND: (Activities on Technical Aspects of Wind Erosion) 168 entries
- PROCOM: UN Compendium on Dryland Development and Desertification Control projects (325 entries)

Other databases:
- DIOR: Directory of Organizations dealing with desertification control and dryland development (537 entries)
- DES-STAT: Graphical representation of world drylands and status of desertification

Table 2. International Soil Reference and Information Centre. ISRIC / UNEP Databases. (www.isric.nl/).

SOTER (World Soils and Terrain Digital Database): Attribute files for soil-terrain maps for selected countries mostly in Latin America, Eastern Europe, the Near East and East Africa; UNEP-ISRIC-FAO collaboration. Information available at scales of 1:1M or smaller; larger scales available in some countries

GLASOD (Global Assessment of human-induced Soil degradation) (1:10M): Produced by UNEP and ISRIC, estimates kind, degree and extent of degradation in all countries. Based on expert judgement of local scientists using agreed upon guidelines.

ASSOD (Assessment of the status of Human-induced Soil Degradation): Regional and national assessments commenced in 17 countries in South and Southeast Asia at scale of 1:5 M. Combines methodologies of SOTER and GLASOD; links soil degradation with crop productivity

SWEAP (SOTER-based Water Erosion Assessment Programme): A SOTER-based programme for assessment of water erosion hazard.

WISE (World Inventory of Soil Emission Potentials): Contains 4353 soil profiles (Africa 1799; South, West, North Asia 522; China, India, Philippines 553; Australia, Pacific islands 122; Europe 492; N. America 226; S. America and Caribbean 599). These profile data are complemented with a simplified grid cell (half degree) database of the World Soil Map

The Core Global Pedon Dataset: 1125 pedons which provide data on the physico-chemical attributes of soils in a homogeneous, internally consistent format for the International Geosphere-Biosphere Programme (IGBP)

Table 3. FAO Databases. (www.apps.fao.org/). FAO maintains a collection of soil and agro-ecological databases, some of which originate solely from FAO, and some jointly between FAO and other international institutions, such as IIASA, UNEP and ISRIC.

FAOSTAT DATA: An online and multilingual database containing over 1 million time series records covering international statistics in Land Use and Irrigation, Production, Forest products and many other areas. FAOSTAT is available on CD-ROM. Also available are databases on:

Agro-ecological Zones Data Bank: This is a global database with data on soils, climate, landforms and some land use

Anticipated (potential) yield: This indicates increasingly intensive land management, signaling the need to check for possible water quality problems and excessive fertilization

CDROM Soil Map of the World: This is a soil database consisting of the digitized version of the Soil Map of the World (1:5M), and various soil interpretations. Geo-referenced soil profiles (descriptions and analytical data) are stored in the FAO/ISRIC Soil Database. It is estimated that FAO maintains records for about 175 soil profiles

The FAO-ISRIC Soil Database (SDB): Derived from the ISRIC Soil Information System (ISIS), contains information on individual soil profiles in coded, numerical, and descriptive format. SDB is a stand-alone programme which does not need a supporting database management system

Table 4. UNEP/GRID/GEMS Related Databases (www.grid.unep.no/).

GEMS/Water and GEMS/Air: Air and water quality monitoring networks

GEO: UNEP's Global Environment Outlook reports. Integrated biennial and decadal outlooks on the global environment; replacing UNEP's traditional state-of-the-environment reports; involving much regional consultation; supported by working groups on data, scenarios, modeling, policy; first issued in 1997

UNEP/GRID Meta-data Directory: See www.grid.unep.no/.

GTOS (The Planned Global Terrestrial Observing System): Comparable and linked to the already existing Global Oceans and Global Climate Observing Systems (GOOS and GCOS). Once operational GTOS will provide an excellent umbrella mechanism for data collection and sharing; co-sponsors are FAO, International Council of Scientific Unions (ICSU), UNESCO, UNEP and World Meteorological Organization (WMO)

Population distribution (through spatial modeling); with the Consultative Group for International Agricultural Research (CGIAR) and the US National Center for Geographic Information (**NCGIA**)

The United Nations Convention to Combat Desertification was negotiated between 1992 and 1994 and came into force on 26 December 1996 when it was ratified by 60 countries. By March 1999 the List of Ratifications shown on the CCD Secretariat's Web site included 145 countries (www.ccd.de). The convention was innovative, offering a negotiated blueprint for environmental management in the drylands with emphasis on the need for community-based action in the field. The first Conference of the

Parties took place in September 1997 in Rome; the second was held in Dakar, Senegal in December 1998.

Table 5. World Bank and other desertification-related Databases. Selected data bases covering soil degradation, desertification, land use and land cover include the following.

LQI (The Land Quality Indicators Programme): (www-esd.worldbank.org/lqi/). The Web Site contains a Meta-database and an inventory of LQI products and a list of LQI Publications. It also gives access links to Global Environmental Data, Bibliographical Catalogues and Modeling information on (www.ciesin.org/lqiis/lqihome.html) run by the **Centre for Earth Science Information Networks (CIESIN)**. The Compendium of Sustainable Development Indicator Initiatives and Publications run by the International Institute for Sustainable Development at (www.iisd.ca/measure/compindex.asp) and the International Development Research Centre (IDRC) (www.idrc.ca/)

CGIAR Centers Databases: Research-oriented databases, maintained by several of the Centers in support of their regional and global research mandates (http://www.cgiar.org)

WOCAT (The World Overview of Conservation Approaches and Technologies): Collects, analyses and distributes knowledge of proven and promising soil and Water Conservation (SWC) practices (www.giub.unibe.ch/cde/projects/wocat.htm)

World Soil Resources Database: Maintained by the USDA – Natural Resource Conservation Service – USDA (formerly the Soil Conservation Service); consists of global, national and regional, digitized soil maps (ARC/Info and GRASS) at various scales, as well as special files on soil pedons (profiles), soil carbon and soil climate (www.statlab.iastate.edu/soils/nsdaf)

Global Land Information System: Maintained by the USGS, contains a wide variety of maps, digital land terrain models, aerial photographs, links to other sites, and other information, clearly catalogued and readily ordered (edcwww.cr.usgs.gov/webglis/glisbin/glismain.pc)

Land cover characterization using Advanced Very High Resolution Radiometer (AVHRR): A joint USGS-EROS Data Centre, UNEP and NASA product, using the International Geosphere-Biosphere Programme (IGBP) processing protocols; Latin America first half 1996; related project implemented for a number of countries in Asia and the Pacific

World Resources Institute Database: This is a PC-compatible database consisting of 503 variables for 198 countries. It is the source database for WRI's publication on global conditions and trends (WRI/UNEP/UNDP/WB). (www.wri.org/wri/index.html). Links to Land Quality Indicators and Sustainable Land Management Indicators site

World Bank Economic and Social Database: The WBESD database contains several million time series in numerous data files from the World Bank, IMF, UN, UNIDO, UNESCO, FAO, OECD, and ILO. These data can be used to develop broad-level, national indicators such as ratios of: Cultivated area/arable land; production/arable land (yield); soil conserving/soil degrading crops; nutrient inputs/nutrient exports. (www.worldbank.org/esd/html/extdr/data.htm)

University of Arizona, Office of Arid Lands Studies: (www.ag.arizona.edu/oals/). Directory of Arid Lands Research Institutions. OALS/FAO/UNEP, 1995

With regard to science, the CCD provides for a Committee on Science and Technology (CST) made up of members appointed by Governments. The CST is focussing on a global network approach to combat desertification, on the establishment of benchmarks and indicators to assess and monitor dryland degradation; on inventories of traditional and local technology, knowledge, know-how and practices, and on the establishment of research priorities. For details on the CST, visit the Secretariat web site (www.unccd.de).

Since the first edition of the World Atlas of Desertification in 1992, efforts to improve the assessment and monitoring of desertification have continued. During the last decade, most countries have prepared national reports: for the Earth Summit; for UNEP's Governing Council; for the Commission on Sustainable Development; for the General Assembly; for the meetings of the Intergovernmental Negotiating Committee; for the CCD and, recently; for the Conference of the Parties to the CCD. These reports are available from the various agencies or national governments. They describe the extent of desertification in each country; the steps being taken or planned; and the continued worsening of the situation in many of the countries. UNDP/UNSO and the CCD Secretariat are now accumulating an important database on the progress of work under the Urgent Action for Africa Programme, and also in other regions, as part of the initial implementation of the CCD.

Although much of this information is of too general a nature to be incorporated into the atlas database, UNEP has continued to encourage and support the gathering and presentation of new and improved data on the drylands. A number of related efforts are under-way at present. Most specific to land degradation is the joint effort between the World Bank, UNEP, UNDP/UNSO, FAO and others on Land Quality Indicators (www-esd.worldbank.org/lqi/). The INCD Secretariat with partner agencies is preparing indicators that can be used to monitor progress of implementation of the CCD. The Commission on Sustainable Development is also working with partners on the preparation of indicators of sustainable development. The latest information on the progress of these programmes can be obtained from the Web Sites listed in Table 5.

2. WORLD ATLAS OF DESERTIFICATION – SECOND EDITION

The first edition of the World Atlas of Desertification was praised by many as an essential reference on the extent of Desertification; but it was also criticized, partly because of the subjective nature of the database and particularly for the absence of references, explanations and sources. As the reprint of the first edition of the atlas sold out, UNEP and the publishers (Edward Arnold) prepared a much revised and improved edition in time for the first Conference of the Parties of the CCD in September 1997 (UNEP, 1997).

In this second edition of the Atlas, the shortcomings of the first edition have been fully addressed. A new, more detailed database, launched at national level in Asia, is described in Section 3. Section 4 of the Atlas is also completely new, with up-to-date coverage of issues and studies on various topics. The explanatory text in Sections 1 and 2 of the Atlas has been expanded and fully revised to reflect new work on GLASOD. All substantive material in the Atlas is fully referenced in a bibliography with almost 700 titles.

Combating desertification involves all aspects of environmental management and is linked with other global concerns such as biodiversity and climate change. The new edition therefore covers a broad range of issues. The availability of water is critical for drylands management and the use of vegetation for household energy has a major impact on the drylands, so these and other issues are covered. Social and economic conditions also have a major impact on the progress and the control of desertification. The new Atlas summarizes the latest information on population (UNDP/UNSO, 1997) and touches on issues such as migration that results from, and leads to, desertification.

3. THE ASSESSMENT DATABASE

Several data sets were used in the compilation of the new edition of the World Atlas of Desertification (Table 2). These include the global soils degradation database, GLASOD; a soils degradation database for continental Africa; a global climatic database and a new, refined soils degradation database "ASSOD" (Assessment of Human-induced Soil Degradation). ASSOD has been developed in South and South East Asia and will be extended world-wide.

The basic indicator of desertification in Sections 1 and 2 of the new Atlas is human-induced soil degradation. Data is available for the global land surface on the types, severity, causes and extent of human-induced soil degradation, at a scale of 1:10 million globally and for Africa at 1:5 million. There are additional data on various other elements important in the drylands equation, such as climatic variability and vegetation degradation. The environment is highly dynamic in the dryland areas, hence it is often very difficult to separate natural processes of land degradation from those resulting from human activities.

The Aridity Zones (the arid, semi-arid and dry sub-humid areas where, by definition, land degradation is desertification), are based on the relationship between precipitation and potential evapotranspiration. This forms an index of moisture deficit termed the "Aridity Index". The maps of climate surfaces, including aridity zones, are derived from the data sets held at the Climatic Research Unit of the University of East Anglia, UK (Hulme and Marsh, 1990). The calculations carried out to determine the Aridity Index and aridity zones are described fully in the Atlas.

Although GLASOD was, by necessity, a somewhat subjective assessment, it was carefully prepared by leading experts in the field. It remains the only global database on the status of human-induced soil degradation, and no other data set comes as close to defining the extent of desertification at the global scale.

To improve GLASOD, a new approach was launched by UNEP with the International Soil Reference and Information Center, the Food and Agriculture Organization, and the relevant national institutions. This commenced in south and southeast Asia. The results are described and displayed in Section 3 of the Atlas. The GLASOD approach has been modified and refined to produce an Assessment of the Status of Human-induced Soil Degradation linking degradation with productivity.

The ASSOD survey was carried out through national institution members of the Asia Network on Problem Soils. Actual degradation assessments were made for 4450 mapping units (compared with 320 for GLASOD for the same area). The classifications include soil characteristics as well as terrain properties, and represent a considerable advance in detail, accuracy and scale over GLASOD. ASSOD has so far involved seventeen countries in south and southeast Asia. Seven of these (China, India, Myanmar, Nepal, Pakistan, Sri Lanka and Thailand) contain areas of susceptible drylands where desertification is an actual or potential problem. It is hoped that the ASSOD work will be extended to cover the globe.

The second edition of the World Atlas of Desertification is a benchmark in a continual process. UNEP's assessment programmes are constantly developing within the limited resources available. Work is carried out by UNEP in house and in collaboration with partner institutions. As assessments improve, UNEP endeavours to ensure that the information is transmitted in appropriate form to the scientific community, the general public and decision-makers.

It is hoped that more comprehensive assessments of the drylands will be possible in time for the tenth anniversary of the Earth Summit in 2002. Now that there is general international agreement on the definition of the problem, it will be possible to determine more precisely its scope.

Land degradation takes many forms and can vary from field to field. It is difficult and expensive to measure at field level over forty percent of the land surface of the globe. Indirect and subjective assessments have been necessary. It is anticipated that new techniques in remote sensing, more widely applied with additional funding resulting from the Convention, will result in new and more precise scientific assessments.

UNEP is continuing to search for new methods and new partners to achieve this in support of the Convention and the Committee on Science and Technology. UNEP is coordinating a consortium of bodies working on the science in order to help the CST draw on the best expertise available. UNEP is also trying to draw more attention to, and achieve better scientific knowledge on, the social and economic factors driving drylands management (e.g., Narjisse, this volume; Sanders, this volume) and to highlight success stories in combating desertification.

4. CONCLUSIONS

The work of revising the World Atlas of Desertification has clarified key inadequacies of the information base and of the scientific understanding and treatment of desertification. It is hoped that these issues will be given serious consideration and that increased support for improving the scientific assessment and monitoring of desertification will be forthcoming. Social aspects of mitigating and alleviating desertification deserve immediate attention. Combating desertification involves achieving sustainable management of the drylands environment so as to meet the needs of the population.

Among the more significant conclusions that have emerged during the preparation of the second edition of the Atlas are the needs:

- To continue with the new ASSOD approach towards the completion of a new global assessment; preferably by the year 2002;
- To increase knowledge and understanding of the social and human dimensions of desertification and on the interactions between physical degradation and social consequences (Stiles, 1995);
- To improve the climatic surfaces and aridity index algorithms to better determine the potential impact of climatic variations on desertification and vice-versa, and to improve the prediction and mitigation of drought (Williams and Balling, 1996);
- To obtain a better understanding of the nature and dynamics of vegetation growth and resilience in the susceptible drylands (Kirkby et al., 1990);

- To improve knowledge of on-site and off-site economic impacts and relationships of desertification (Dixon et al., 1988);
- To improve the knowledge of the human migration issue (e.g., Westing, 1994; Schwartz et al., 1995) and its causes and effects;
- To broaden the approach to assessing and monitoring change in the susceptible drylands to incorporate additional characteristics including water, social and economic indicators;
- To record, respect, evaluate and develop traditional knowledge and traditional practices (Barraclough, 1995);
- To carry out further research on carbon storage and on the feasibility of strategies to enhance carbon sequestration (Squires et al., 1998);
- To encourage throughout the drylands the practice of sustainable management to ensure that the long-term needs of the land are respected, at the same time as meeting immediate productivity goals (Behnke and Scoones, 1993);
- To continue to investigate, develop and use new methods of monitoring that would enable improved scientific assessment of land degradation at field level (Warren and Agnew, 1988).

ACKNOWLEDGEMENTS

The author wishes to acknowledge the assistance of Dr. Till Darnhofer, Dr. Ali Ayoub, Dr. Johannes Akiwumi and other colleagues at UNEP in the preparation of this document.

REFERENCES

Barraclough, S. 1995. Social dimensions of desertification: a review of key issues. In: Social Aspects of Sustainable Dryland Management. Ed. D. Stiles. pp. 21–79. Wiley, Chichester.

Behnke, R.H. and Scoones, I. 1993. Rethinking range ecology: implications for range management in Africa. In: Range Ecology at Disequilibrium. Eds. R.H. Behnke, I. Scoones and C. Kerven. pp. 1–30. Overseas Development Institute, London.

Dixon, John A., James, D.E. and Sherman, P.B. 1988. The Economics of Dryland Management. IIED, London.

FAO and UNEP 1984. Provisional Methodology for Assessment and Mapping of Desertification. Rome.

Hulme, M. and Marsh, R. 1990. Global mean monthly humidity surfaces for 1930–59, 1960–89 and projected for 2030. Report to UNEP/GEMS/GRID. Climatic Research Unit, University of East Anglia, Norwich.

Kirkby, M.J., Atkinson, K. and Lockwood, J. 1990. The interaction of erosional and vegetational dynamics in land degradation: spatial outcomes. In: Vegetation and Geomorphology. pp. 25–40. Wiley, Chichester.

Schwartz, M. Leighton and Notini, J. 1995. Desertification and Migration: Mexico and the United States. US Commission on Immigration Reform, Washington DC.

Squires, V.R., Glenn, E.P. and Ayoub, A.T. 1998. Land Management, Desertification and the Global Carbon Cycle. University of Arizona Press, Tucson, AZ.

Stiles, D. (ed.). 1995. Social Aspects of Sustainable Dryland Management. Wiley, Chichester.

United Nations 1978. United Nations Conference on Desertification. Round-up, Plan of Action and Resolutions. New York.

United Nations 1994. United Nations Convention to Combat Desertification in those Countries Experiencing Serious Drought and/or Desertification, Particularly in Africa. United Nations, New York.

UNCED 1992. Earth Summit, Rio de Janeiro. Regency Press, London.

UNCED 1992. Agenda 21. United Nations, New York.

UNDP/UNSO 1997. Aridity Zones and Dryland Populations: an Assessment of Population Levels in the World's Drylands with Particular Reference to Africa. UNDP Office to Combat Desertification and Drought (UNSO), New York.

UNEP 1992a. World Atlas of Desertification. Edward Arnold, London.

UNEP 1992b. Status of Desertification and Implementation of the United Nations Plan of Action to Combat Desertification. Nairobi.

UNEP 1997. World Atlas of Desertification. 2nd edn. Eds. M. Thomas and N. Middleton. Arnold, London & Wiley, New York.

Warren, A. and Agnew, C. 1988. An assessment of desertification and land degradation in arid and semi-arid areas. Paper 2. IIED, London.

Westing, A.H. 1994. Population, desertification and migration. Environmental Conservation 21, 110–114.

Williams, M.A.J. and Balling, R.C. 1996. Interactions Between Desertification and Climate. Edward Arnold, London.

NOTE ON INTERNET ADDRESSES

Please note that, while the World Wide Web addresses provided were correct in March 1999, they are likely to change from time to time.

The implementation of soil conservation programmes

David Sanders[1]

Flat No. 1, Queen Quay, Welsh Back, Bristol BS1 4SL, UK

Tel&Fax: 44 117 9276021; E-mail: <sanders@clara.net>

ABSTRACT The problem of land degradation has worsened in recent years in spite of the vast amounts of money and effort that have gone into conservation programmes. This paper looks at reasons why programmes have not been more effective and attempts to identify the factors that must be addressed if soil conservation programmes are to be more successful. It is pointed out that many past programmes have failed because they have concentrated on the symptoms of the problem rather than the underlying causes. This means that such issues as land tenure and access to markets often have to be addressed before any real progress can be made. It is now recognized that large-scale changes in the way land is managed and used depends upon the perceptions of the many individual people who use it. Soil conservation programmes can therefore only hope to succeed if these people are fully involved in the whole process of identification of the problems, developing and then implementing solutions. While participation is important, landusers will not change the way in which they manage land unless the right blend of incentives and disincentives exists. Some of the incentives that affect soil conservation programmes are discussed, while particular attention is paid to the subjects of land tenure, the development of appropriate technology, research-extension-landuser links, the farm family and gender issues, continuity of programmes and monitoring and evaluation.

Key words: conservation incentives, extension, gender issues, land tenure, participation, problem identification, rangeland desertification.

1. INTRODUCTION

Land degradation is an old problem which has faced mankind ever since land was first settled and cultivated some 7,000 years ago. Over the centuries people have developed effective strategies and techniques to protect and rehabilitate the land. The remains of what has been done in the past can still be found in the old terracing systems in Yemen, China and Peru, as well as in

---

[1] David Sanders – formerly Senior Soil Conservation Officer with the Food and Agriculture Organization of the United Nations, now President of the World Association of Soil and Water Conservation.

traditional farming systems such as shifting cultivation, which is still practised in parts of the tropics, and nomadic grazing in the semi-arid regions of Africa, Asia and the Middle East (Sanders, 1990). Clearly, farmers have understood the principles of soil conservation for centuries.

Large-scale, modern soil conservation programmes started in the United States in the 1930s. At this time soil erosion was becoming a serious problem in other parts of the world and large-scale programmes were soon to start up in Africa, Australia and, a little later, in India and other Asian countries. Since that time, dozens of programmes and projects have been implemented and billions of dollars have been spent on soil conservation. In spite of this, the problem of land degradation is still very much with us and in many countries the problem is getting worse rather than better.

If the principles of soil conservation have been understood for centuries and a vast amount of time, effort and money has gone into soil conservation programmes over the last 60 years, why is soil erosion still such a serious problem? The answer is complex, but we can identify a number of interrelated factors from the experiences of past and ongoing programmes. This chapter addresses these factors.

2. IDENTIFYING THE REAL CAUSES OF THE PROBLEM

A major reason for the failure of many soil conservation programmes is that they have not tackled the real causes of the problem. Land degradation, in such forms as soil erosion, salinization and fertility decline, is but the physical manifestation of land being used in the wrong way and being badly managed. In most parts of the world this has not been understood and programmes have concentrated on dealing with the symptoms, not the real problems. For example, a range reseeding programme may have no effect without also resolving the problem of overgrazing. Yet, how often have we seen programmes of revegetation and "regreening" attempted in the Middle East and North Africa without any real effort being made to overcome the problems of overgrazing and deforestation which led to the problem?

Farmers, graziers, foresters and other landusers depend on the productivity of the land for their livelihood. They do not intentionally encourage it to degrade. Incorrect land use and bad management primarily result from economic, social and political pressures which force them to use land in the way that they do (FAO, 1990; Narjisse, this volume).

The first step in developing any successful conservation programme should therefore be to identify what undesirable land uses are being practised and then to analyse why. Reasons may include population pressures, inappropriate agricultural pricing policies, unavailability of inputs, or land tenure systems that force people to over-exploit the land (Narjisse, this volume). Without analysis, the underlying causes of land degradation may well be overlooked, while much time, effort and money is spent on dealing with symptoms.

Many examples exist where this has happened. In an attempt to increase agricultural production, governments frequently introduce measures which lead to land degradation. For example, several Middle Eastern and North African countries have, at various times, heavily subsidised the price of grain for stock feed. While this has temporarily achieved the object of increasing meat production, it has led to a large build up of sheep numbers and to extensive damage to the region's rangelands. Programmes which concentrate on the physical manifestations of the problem, such as rangeland reseeding and water spreading, are bound to fail unless something is done to overcome the fundamental problem of how animals are managed.

3. FINDING THE SOLUTIONS

3.1 People's participation

Early conservation schemes concentrated on technology. Land degradation was seen as a physical problem and it was believed that the answers lay in technology that had to be developed by research workers. Once perfected on the research station, the task was simply to pass the techniques on to the landusers for implementation. The landusers were usually seen as the recipients of the process and were not expected to contribute other than to install and maintain the advocated works (Griesbach and Sanders, 1996).

It has since been shown that this approach does not work (FAO, 1990). Firstly, it overlooked the fact that land degradation is only the symptom of the problem – the physical manifestation of the land being incorrectly used because of economic, social, political and legal pressures. Secondly, the solutions offered were often unattractive to the landusers as they usually involved additional work, but did little to solve the immediate problems of improving yields, increasing income or lessening risks. As a result, the technical solutions offered were often seen as irrelevant.

In the 1980s, the approach to soil conservation began to change in many countries. It became widely recognized that the way a country's land was managed and used depended upon the perception of its individual landusers. These people have the ability to bring about fundamental changes in land use for the better. It was seen that for this to happen the people themselves had to be far more closely involved in the process of identifying the problems, working out the solutions and then implementing what needed to be done. This has led to great changes, with far more effort being made to involve the landusers (A. Arnalds, this volume).

To date, the results have been mixed for various reasons. "People's Participation" has become very fashionable with donors. To attract funding, many projects labelled "participatory" were only participatory in name. In addition, it has been discovered that to achieve true and effective participation of landusers can be time-consuming and difficult. New attitudes and special skills, not present in the staff of most conservation agencies, may be required (Hagmann et al., 1996).

Nevertheless, a number of good examples have shown that full involvement of the landusers can lead to popular and effective programmes. The largest, and arguably the most successful, is the Landcare Program in Australia. A feature of this programme has been the assumption of managerial responsibilities for all phases of the programme, from planning to implementation, by the landowners themselves. So effective has this programme been since its launch in the mid-1980s, that more than a quarter of the country's farmers have participated (Campbell, 1994).

3.2 Creating the right blend of incentives and disincentives

Even if the underlying reasons for land degradation are identified, the landusers fully involved and possible solutions identified, soil conservation programmes may still not function effectively because the solutions may not be sufficiently attractive for landusers to adopt, for strong social, economic or institutional reasons. In most cases incentives or disincentives are needed, in one form or another, before landusers will change their ways of using and managing the land.

Just as people's participation was identified as an important prerequisite for successful conservation programmes in the 1980s, the importance of incentives and disincentives is now being recognized in the 1990s. At the time of writing, at least one major agricultural funding agency, the International Fund for Agricultural Development (IFAD), was studying the use of incentives in conservation programmes (Sourang, 1996). The World Association of Soil and Water Conservation (WASWC) is in the process of publishing a book on the subject.

Over the years, a wide variety of incentives and disincentives have been developed by government and donor agencies for conservation programmes. These exist in most countries in one form or another. Incentives (and disincentives) can be divided into two main categories, direct and indirect.

Indirect incentives are the most powerful and are likely to have the most profound effect. After reviewing a study for IFAD on this subject, Sourang (1996) concluded that the importance of indirect incentives cannot be overestimated, particularly land tenure and user rights, markets and prices and decentralization of decision making. This underlines the important influence of government policy and macro-economics on land use. Some of these factors are dealt with below; but first the need for direct incentives is examined.

Direct incentives can be effective, if well thought out. This has proved to be the case with the Landcare Program in Australia, where a number of incentives, such as small grants, can be obtained under certain circumstances. However, there are many examples of direct incentives having had very little long-term effect, with landusers being prepared to carry out prescribed works only for the sake of obtaining a short-term reward, then quickly reverting to their old ways once the incentive had been used or withdrawn. For example, in the 1970s and 1980s massive quantities of food were distributed to farmers in Ethiopia in payment for the completion of erosion control works, such as the construction of earth and stone contour banks and tree planting. After some years, it was discovered that the farmers were not interested in the erosion control works, but only in receiving the food rations. After the food was received, the contour works were allowed to break down or, in some cases, were deliberately destroyed so that the farmers could be paid to build them again. In addition, trees were cut down once protective policing was removed (Sanders and Cahill, 1998).

Incentives, whether direct or indirect, should be relevant to "farmer pull", rather than "technology push" (De Graaff, 1993). In other words, if incentives are to be effective in the long run, they must be orientated towards the problems as perceived by the landusers, instead of being focussed on the wide-scale implementation of technical measures whose relevance the landusers do not understand or do not have the resources to maintain.

3.3 Appreciating the importance of the land tenure system

It is now recognized that land tenure plays an important part in how well land is managed and conserved. Farmers with no long-term land rights or access to land are unlikely to invest in improvements or labour intensive works like terracing or tree planting. If landusers are granted long-term rights of use, the position can change dramatically.

Wise adjustments by governments to land tenure systems can provide some of the most effective indirect incentives for landusers to practise better land management. How an improved form of land tenure can be applied to the vast rangelands, which cover much of the world's developing countries, still remains to be resolved.

3.4 The importance of technology

With the emphasis on people's participation and landusers developing their own solutions to problems, there is a tendency to forget the importance of technology and the need for research and the development of new practices. As social and economic conditions change, technology needs to change. Conservation programmes will not succeed unless the right technology is available.

The fact that many traditional farming systems contain very effective conservation practices has become recognized in recent years. These can be seen in the old terracing systems of Yemen, in nomadic grazing systems in the Middle East and shifting cultivation in tropical Africa.

Farming systems, particularly small-scale and subsistence systems, tend to be complex. It is often very difficult to integrate completely new practices into them. However, there are usually possibilities of adapting and improving the traditional systems so that they can once more be used effectively. This is because the underlying principles are already known and understood and the landusers have the skills to put them into practice. This was demonstrated in the 1960s when efforts were made to introduce soil conservation practices into the badly eroded dryland areas of Jordan. A number of different practices were tried but none of those based on technology developed in other countries was adopted by the farmers. However, one based on the traditional practices of stone wall terracing was readily taken up. In this case, the farmers were shown how to align the walls on a true contour and use improved construction methods. The wall slowed runoff, increased infiltration of water into the soil, caught silt and slowly developed into terraces. Barren land was made fertile. Since then, thousands of hectares have been treated in this way on hundreds of individual farms. The lesson here is that technology should be built, where possible, on locally known practices and skills (Sanders, 1988; A. Arnalds, this volume; Narjisse, this volume).

3.5 The need for research-extension-landuser links

The development of soil conservation technology must be relevant to the perceived need of the landusers. In the past this has not always been the case. With changing approaches to soil conservation, the relationship between the landuser, extension worker and researcher is changing. No longer can the relationship be simply that of new technology being passed down from researcher, through extension, to the farmer. Soil conservation still needs research stations and some of the traditional types of research, but the demand is now for a new relationship under which the problems are first defined by consultations between the three parties who then work together to develop solutions. In practice, this means extension workers and landusers taking an active part in research, with more on-farm trials and the development of a close dialogue between all the parties concerned (Onchere, this volume).

There are a number of good examples where this is now being done. For example, the International Board for Soil Research and Management (IBSRAM) is assisting national research centres in Asia to test and develop different technologies, not only for their technical soundness, but also for their acceptability to farmers – something that has often been neglected in the past.

3.6 The importance of the farm family and gender issues

An integral part of the farm is the farm family. For conservation programmes to succeed, the operation and capabilities of those running the farm must be considered. Gender issues are extremely important. In many rural communities the women do most of the farm work. This fact is usually overlooked and most extension workers are men who tend to work with the male members of farm families. As a result, schemes are often developed which cannot work because they do not take into consideration the capabilities and responsibilities of women. For example, in many parts of Africa women spend many hours each day carrying water as part of their duties. This, with their household and farming responsibilities, leaves them with little time for anything else. Under these conditions, it may be unrealistic to expect them to start labour intensive activities, for example the stall feeding of cattle. The first step may be to provide a better water supply to free up time to devote to other, more productive activities. More female extension workers are needed if problems such as these are to be recognized and dealt with (Onchere, this volume). This is particularly so in places where the culture does not allow male extension workers to work directly with women.

3.7 Stability and continuity

Continuity is an important element of successful soil conservation programmes. Short-term programmes have seldom achieved much. The reason for this is that soil conservation practices are only accepted and adopted by landusers after what is frequently a long process of developing and testing different alternatives. New practices have to be seen to work successfully under different conditions, people have to learn new skills and attitudes and behavioural patterns have to change. This process is usually slow and may take many years. For these reasons, programmes are most likely to succeed if they start on a small scale and gradually build in size over the years, as experience and knowledge are gained and as a trusting relationship is built up with the landuser.

3.8 Monitoring and evaluation

Few soil conservation programmes have given enough attention to monitoring and evaluation. This is partly due to the problem of defining clearly what exactly is meant by soil and water conservation, particularly now that the emphasis is being placed more on productive measures and the diversity of soil conservation practices that can be found. Effectiveness can no longer simply be measured in terms of a reduction in the number of tonnes of soil lost. Nevertheless, a number of methods for monitoring and evaluating do exist, ranging from the use of aerial photos to baseline surveys of sample areas (see Tongway and Hindley, this volume; Imeson and Cammeraat, this volume). Certainly the landusers should be closely involved in the monitoring and evaluation process as their perceptions of what is successful may not be the same as those of research or extension workers (A. Arnalds, this volume). Until monitoring and evaluation are used in a more systematic way, it will be difficult to modify and improve programmes as they proceed and prevent the repetition of many of the mistakes that we see being repeated in different parts of the world.

4. CONCLUSIONS

Most conservation schemes in the past have based their programmes on combatting the physical effects of land degradation, with little thought being given to the underlying causes. As a result, programmes have generally dealt with the symptoms rather than the real problems. Not until more time and effort are spent on dealing with the real causes of land degradation will there be an improvement in the performance of conservation projects.

Nevertheless, considerable progress has been made since the early 1980s as more attention has been given to people's participation in the whole process, from problem identification, through developing solutions, to implementation and maintenance of the required measures. The importance of incentives and disincentives is only now being recognized and we can expect to hear a great deal more on this subject over the next decade. It has already become clear that direct incentives are difficult to apply successfully and there are many examples of where these have proved to be ineffective and even counterproductive in some cases. On the other hand, indirect incentives are proving to be very important in deciding whether or not landusers will adopt more conservation effective systems of land use and management. Changes in land tenure systems, in particular, are proving to have a very important effect on land management.

With the present emphasis on socio-economics, it is easy to overlook the importance of technology. However, conservation programmes cannot be expected to succeed unless appropriate technology is readily available to the landusers. In relation to this, there is a need for a closer relationship between researchers, extension workers and landusers in most countries. In the past, not enough attention has been paid to the capabilities and limitations of the farm family, particularly the female members of the family, who are already committed to very heavy workloads in many countries.

Finally, programmes need to be long-term to allow landusers to test, evaluate and adopt new practices. There is need for constant monitoring and evaluation to avoid repeating the mistakes of the past.

REFERENCES

Campbell, A. 1994. Landcare – Communities Shaping the Land and the Future. Southwood Press Pty. Ltd, Sydney. ISBN 1 86373 555 0.

De Graaff, J. 1993. Soil Conservation and Sustainable Land Use – an Economic Approach. Royal Tropical Institute, The Netherlands. ISBN 90 6832 042 4.

FAO. 1990. The Conservation and Reclamation of African Lands – an International Scheme. The Food and Agriculture Organization of the United Nations. ARC/90/4, Rome. W/Z5700E/3/2.93/1000.

Hagmann, J., Murwira, K. and Chuma, E. 1996. Participatory development and extension of soil and water conservation in southern Zimbabwe. In: Soil Conservation Extension – from Concepts to Adoption. Eds. Sombatpanit, Zobisch, Sanders and Cook. Soil and Water Conservation Society of Thailand. ISBN 974-7721-70-8.

Griesbach, J-C. and Sanders, D.W. 1996. Soil and water conservation strategies at regional, sub-regional and national levels. In: Towards Sustainable Land Use, Advances in Geoecology 31. Vol. II. Catena Verlag GmbH, Reiskirchen, Germany.

Sanders, D.W. and Cahill, D. 1998.Where incentives fit in soil conservation programmes. In: Using Incentives in Soil Conservation. Eds. Sanders, Sombatpanit, Huszar and Enters. Oxford & IBH Publishing Co. Pvt. Ltd., New Delhi. (In press).

Sanders, D.W. 1988. Food and agriculture organization activities. In: Soil Conservation in Conservation Farming on Steep Lands. Eds. Moldenhauer and Hudson. Soil and Water Conservation Society, Ankeny, Iowa. ISBN 0-935734-19-8.

Sanders, D.W. 1990. New strategies for soil conservation. The Journal of Soil and Water Conservation 45(5).

Sourang, CM. 1996. Incentive systems for natural resource management. A paper presented on the occasion of the International Forum on Local Area Development under the Convention to Combat Desertification, IFAD, Rome.

Evolution of rangeland conservation strategies

Andres Arnalds
Soil Conservation Service, Gunnarsholt, IS-850 Hella, Iceland
Tel: 354 477 5500; Fax: 354 487 5500; E-mail: <andres@landgr.is>

ABSTRACT The history of soil conservation approaches worldwide may be characterized by a change from single issue, top down approaches to ecosystem management and community involvement. In this paper, the long history of combating desertification in Iceland is used to illustrate past, present and future perspectives in the implementation of rangeland conservation programs.

Livestock production based on grazing of rangelands has been the mainstay of Icelandic agriculture. Unsustainable land use for the past 1100 years has, however, led to severe degradation. Tree-cover has declined by about 96% and much of the original vegetation cover has been lost. The composition of remaining vegetation varies, but long term over-grazing has led to a low proportion of palatable species in many areas.

After 90 years of soil conservation in Iceland, approaches have been undergoing rapid changes. Past programs emphasized federal intervention, agronomic approaches (seeding, fertilization) and degradation "containment". As a consequence, soil conservation came to be regarded as a government responsibility, rather than a responsibility of the land user. In contrast, current programs emphasize sustainable land use, land improvements, environmental quality, land literacy, and locally led participation. More than 25% of the sheep farmers and thousands of volunteers currently work with the national Soil Conservation Service on restoration projects and combating rangeland desertification. Their direct involvement, combined with various other incentives, has been a powerful tool in fostering a "conservation ethic".

Key words: carbon sequestration, conservation incentives, land ethic, land literacy, participatory approaches, rangeland desertification, soil conservation, sustainability.

1. RANGELAND DESERTIFICATION

Although desertification, as defined in the context of the Convention to Combat Desertification, is confined to the arid lands of the world, it is a problem of global nature (O. Arnalds, this volume). The history of the soils and vegetation resources of Iceland is a classic example of the long-term consequences of the human battle for survival in a rigorous and sensitive environment (Gisladottir and Preston-White, 1998).

The settlers who came to Iceland 1100 years ago (about 874 AD) saw a fertile land. Vegetation may have covered more than 60% of the country, and woodlands, mainly birch (*Betula pubescens*), covered at least 25% of the land area. The vegetative cover provided good protection of fragile volcanic soils. With the settlement, a delicate balance between a hostile climatic environment and vulnerable vegetation was disrupted (Thorsteinsson, 1986; Arnalds, 1987). The woodlands were cut for fuel or burned to make pastures. High grazing pressure damaged the land and interfered with vegetation recovery after natural disturbances. Subsequent soil erosion has devastated large parts of the ecosystem, reducing vegetative cover by an estimated 50%. Trees now cover only 1% of the land area.

Desertification continues to be a major threat to Iceland's natural resources. A national assessment of soil erosion indicates that 40% of Iceland is experiencing severe soil erosion (Arnalds et al., 1997). This massive ecosystem degeneration entails high costs for Icelandic society, including reduced agricultural productivity and food security, loss of shelter from frequent high winds, and degraded watershed hydrology. Mitigating poor rangeland health and continued soil erosion are among Iceland's highest priority environmental issues.

2. FROM SINGLE-ISSUE SOIL CONSERVATION TO ECOSYSTEM MANAGEMENT

With one of the oldest soil conservation agencies in the world, Iceland has a long history of successes and failures in combating rangeland desertification. This experience has parallels in many parts of the world. Most of the 90 years of state organized soil conservation in Iceland have been characterized by localized efforts to contain the spread of catastrophic soil erosion. In contrast, the last decade has been characterized by an emphasis of natural resource conservation based on management for sustainability.

2.1 The first phase of soil conservation in Iceland

Catastrophic soil erosion has been occurring in many parts of Iceland for centuries (Arnalds, 1987). One of the worst periods of degradation may have taken place in the late 19th and early 20th century; a consequence of an increase in livestock numbers interacting with frequent cold spells which diminished plant production. During this period, erosion, especially sandstorms, brought destruction to large areas, resulting in the abandonment of many farms.

An organized battle against desertification began in 1907 with an Act of Parliament and the founding of the Icelandic Soil Conservation Service (SCS). The first 50 years were primarily devoted to fighting catastrophic soil erosion and rapid sand encroachment. Fences were erected to protect areas from grazing, and Beach rye (*Leymus arenarius*) was seeded, as it was the only plant capable of successfully establishing and binding moving sand in Iceland. Stone or timber walls were also built to shelter plants.

Much was achieved and towns or farms are no longer threatened by sand encroachment. In recent decades emphasis has shifted to restoring some of the lost resources, and vegetation is being re-established in many areas.

2.2 Lessons learned

Despite many victories in the long history of soil conservation, soil erosion is still an acute problem. In large areas plant cover and condition is not adequate for protection against soil erosion. The health of most of the Icelandic rangelands is far from being adequate and unsustainable land use practices continue to threaten natural resources.The plant communities that characterize degraded ecosystems are of low productivity and do not represent site potential.

Approaches applied during the first phases of soil conservation mirrored those in many other parts of the world (Breckwoldt, 1988; Sanders, 1992 & this volume). The focus was on single-aspect soil conservation with low local involvement. The traditional approach involved using governmental personnel and equipment to halt localized catastrophic soil erosion. As a consequence, soil conservation came to be regarded as a governmental responsibility, rather than an ethical obligation of the land user. Soil conservation fences, designed to protect land from grazing, were primarily erected at sites of catastrophic erosion, without consideration of land condition at larger scales. This practice may have intensified off-site impacts, as livestock was simply moved to non-excluded areas. This "Band-Aid" approach dealt primarily with the symptoms of degradation rather than the causes.

2.3 New approaches

New approaches to soil conservation, developed over the past decade, are based on improved resource information, analysis of barriers to conservation and targeted incentives. The most important tools in maintaining or improving ecosystem quality are considered to be increased participation and responsibility of land users, and indeed the whole population. Various incentives for soil conservation are being developed. Most important of these has been a gradual shift away from a narrow focus on agricultural production without regard for environmental quality, towards sustainable land use (Arnalds, 1999).

Methodologies in rangeland conservation are also changing. Research is leading the way to a shift from an emphasis on seeding and fertilization towards ecosystem management based on the principles of restoration ecology.

3. THE QUEST FOR SUSTAINABLE LIVESTOCK PRODUCTION

Agriculture in Iceland basically means pasture- and rangeland-based livestock production, as the climate does not allow for large-scale grain production. Farms and all urban areas are located in the coastal lowlands, and most farms are privately owned. Grazing continues to be a major determinant of the health of Iceland's ecosystems. The vast interior highlands are used as "commons" for summer grazing by sheep, in communal districts. Farmers own or have grazing rights to most of Iceland, but the ecological condition of most of the communal grazing areas is very poor. There is an urgent need to minimize erosion and enhance vegetation succession on much of this land.

The SCS further aims at full protection of the most degraded highland commons from grazing and a sharper focus on sustainable use on the remaining grazing lands.

Restoration of degraded land and the quest for sustainability is unattainable without a management commitment by the agricultural community. A conservation barrier, resulting from the long era of "top down approaches" had to be overcome. It had left the farmers with a poor perception of the degradation problems, resulting in a defensive attitude and minimal responsibility for solving those problems. Contemporary approaches focus on creating awareness by increased participation in soil conservation projects, with a local leadership emphasis. The long-term goal is to make the land users the true custodians of the land.

The direct involvement has proved to be a very powerful incentive, and the results are similar to the experience of the Australian Landcare Program (Campbell, 1994), the US Environmental Quality Incentives Program (Molleur and Loser, 1998), and other programs utilizing the power of grass-root approaches (Hurni, 1996; Sanders, this volume).

3.1 Farmers reclaim the land

Programs that provide livestock owners with voluntary incentives to adopt sound conservation practices have been evolving. The most extensive is the *Farmers Reclaim the Land* program, centered on assisting landowners to develop and reach their own reclamation goals. Profitability of sheep farming in Iceland is dependent on rangeland quality (Arnalds and Rittenhouse, 1986). This economic link has been an incentive to many farmers to participate in the *Farmers Reclaim the Land* project.

Realizing the limited financial resources for land improvement, it was decided that farmers' machinery, time and skill, plus 15% of fertilizer costs would be their main contribution to the project. The direct cost-share, although limited, has greatly stimulated the feeling of "ownership" of the project. The project was built on the psychological concept of mutual trust. Bureaucracy was minimized and good communication; a handshake and simple paper work was emphasized.

The "bottom-up" nature of the project has made it easy to interact with land users on their own terms. The farmers take pride in their achievements and enjoy being participants in reaching a solution to degradation problems. This, in turn, opens up positive channels for discussing and resolving other resource issues; topics that were traditionally difficult to broach prior to this co-operation. The farmers are innovative and have been urged to experiment with reclamation guidelines and customize solutions for their situations. This farmer–scientist co-operation has greatly advanced the development of successful restoration techniques, and illustrates the power of the grass-root approach. More than 25% of the sheep farmers in Iceland are now actively participating, along with a number of horse owners and dairy farmers.

The *Farmers Reclaim the Land* program is focused on eroded land and the chronic forms of soil erosion. Fighting catastrophic erosion is beyond the scope of this project, and this task is a legal responsibility of the government. However, the organization of this work has been changed, so that local people, especially farmers, now carry out the majority of such projects under the supervision of SCS staff. For the farmer, this means added income in tough times. For the SCS, the benefits include high quality work, a substantial cost reduction, and an attitude change among the farmers who now develop a feeling of "ownership" in the results.

3.2 Conservation and land use planning

The ability to plan land use is one of the keys to sustainability. A decision to make detailed maps of all farms in Iceland over the next decade was recently made. This will be a joint project of several partners under the leadership of the Agricultural Research Institute, SCS, Forestry Service, and the Farmers Union. The maps, based on both satellite imagery and aerial photography, will show constructions and land characteristics (vegetation, topography, soils, and land capability), give assessments of suitability to grazing, and point out restoration needs.

Care will be taken to bring the localized phase of the planning to the grass-root level. The process of making such plans should be one of facilitation, utilizing fully its power as a tool for improving land literacy and increasing conservation awareness. This further assures farmers "ownership" in the results, which in turn stimulates them to adopt the plans. It is anticipated that governmental support for land reclamation will be linked to conservation plans in order to maximize efficiency and to satisfy taxpayer concerns. Development of such plans may eventually form a basis for reciprocal farmer-government commitments with regard to agricultural support.

3.3 Market incentives and governmental support

With growing environmental concern in the world, livestock producers see potential for marketing "environmental friendly" products. Standards are now being developed making it possible to trace the product from the range to the market. Rangeland health and quality of management will be a part of those standards. Participation will be voluntary. It is anticipated that the verification process will be a powerful educational tool, and it is hoped that "eco-labels" will be helpful in the creation of conservation ethos targeting problems as well as solutions.

A relatively high level of government support has characterized agriculture in Iceland. Such support has been granted without regard to land condition. The government may thus be regarded as maintaining, or even intensifying, land degradation in many areas. Fearing tax-payer protests, many farmers are now realizing that support for maintaining the livelihood of the farmer may come to be regarded as a social contract, and that progressive management will be required for its maintenance (Arnalds, 1999).

3.4 Horses – a growing management concern

Currently there are 82,000 horses in Iceland. This population has an estimated doubling time of about 25 years, based on trends over the last 10 years. Horses are grazed almost entirely on private land, and are owned by farmers and by urban dwellers who lease or own land. The development of conservation awareness has been particularly difficult among horse owners. They are a heterogeneous group, spanning a wide spectrum of society, and their concerns have been more related to breeding than the ecological condition of the land.

A survey of overgrazing problems was one of the first steps towards overcoming the traditional denial stage. The results indicated frequent overgrazing, in many cases resulting in severe land degeneration. These alarming findings were widely publicized in both horse-journals and the general media. The open discussion has greatly assisted in the development of much-needed awareness, resulting in public pressure and in action by the horse owners in response to the management problems.

Soil conservation authorities and leaders of the horse owner groups are now seeking joint solutions to attain sustainability. These include increased advice on horse grazing management, educational campaigns, and voluntary reductions in the number of horses. Government taxation on horses has been suggested. The most powerful incentive, in the long run, may be a quality control scheme, that is being developed by the "horse industry". It is intended to give a marketing edge on increasingly quality conscious markets. Included in the verification of production quality will be sustainability considerations and independent judgements of land condition. A direct link between the land and environmentally conscious buyers is expected to lead to improved management, whereas peer-pressure is likely to aid in resolving management problems.

4. CARBON SEQUESTRATION – A POWERFUL INCENTIVE FOR RESTORING RANGELAND HEALTH

The degraded rangelands of the world have a great potential for carbon sequestration both in soil and biota, providing for long-term storage of atmospheric carbon. Reducing release of stored carbon to the atmosphere is of major importance in achieving stabilization of green house gas concentrations and mitigating climate change (Arnalds et al., 1999).

Restoring ecosystem fertility and reducing atmospheric green house gas accumulation can be integrated to the benefit of all. Inspired by the long history of successful soil conservation and forestry in Iceland, the government decided in 1995 to include carbon sequestration as an important part of the National Climate Change Action Program for the period 1990–2000. The five-year special program resulted in a 30% increase in the overall budget for soil conservation and forestry activities in Iceland. Linkage of this program with the goals of the Conventions of Climate and Desertification has become one of the main financial incentives for increasing government funding of soil conservation and land reclamation programs.

Currently, only forestry issues are accepted in the Kyoto protocols. Formal approval of soil carbon sinks for the next period might be a powerful incentive for both government and industry funding for conservation and restoration of degraded land. Small trees and shrubs often characterize rangelands. Recognition of these characteristics is important with regard to "forest" definitions in the Kyoto protocols. Broadening of that definition by including shrublands and woodlands would greatly encourage the use of carbon sequestration as an incentive for combating desertification and improving rangelands in vast areas of the world, including Iceland. The "Clean Development Mechanism" of the Kyoto Protocol may further enhance such a "win-win" situation, as industries needing carbon credits may help pay for landcare programs that might otherwise not be financially feasible.

5. CONSERVATION ETHIC

The creation of a conservation ethic, or land ethic, is one of the main goals of any soil conservation work (Roberts, 1986; Roberts, 1989). Actions are influenced by attitudes, which in turn reflect our state of knowledge. An active SCS role in environmental education is therefore essential, in the school system, with the farmers, and with all of those who affect land condition directly or indirectly. Every land user must recognize the effects his actions have; and his responsibility must also be clear. Such awareness requires a thorough understanding of ecosystems, land use effects, and means to restore damaged land.

Although public opinion polls indicate that most Icelanders acknowledge soil erosion to be a major environmental problem, their actual perception of the health of the rangelands has been questionable. The first national survey of soil erosion in Iceland (Arnalds et al., 1997) confirmed the severity of this problem. Linked to the survey was a land literacy campaign aimed at land users, schools, and the general public. This project received the Nordic Council Environmental Award in 1998. The result has been a much needed change from the problem-denial syndrome that often characterizes the "we think" state of knowledge. Such problem recognition has had wide-ranging effects. At the political level, the survey results formed the foundation for a new soil conservation program. As the voice of the general public calling for action becomes louder, land users will become increasingly responsible for their management and more active in restoring rangeland health.

The media (e.g., newspapers, TV, radio) has played a significant role in improving conservation awareness, working at times closely with SCS personnel. In such co-operation, it is important to reach a balance between problems and solutions. Instead of stimulating actions aimed at resolving the problems, too much focus on erosion and land-use problems can lead to pessimism. Most of the damage to the rangelands may have been done by past generations, but in the media generalizations, current land-users often take the blame. As a result, there was a period of antagonism against conservation characterized by a defensive "I don't care" attitude.

Conservation efforts must also be ecosystem oriented. The emphasis on the visually striking soil erosion in Iceland has resulted in a skewed picture of rangeland health. The large proportion of low productivity, unpalatable plant species dominating many rangelands in Iceland, resulting from past abuse of the land, has received little attention. Most Icelanders are not aware of this issue and its relationship to ecosystem stability and sustainability. Most people also find it difficult to visualize the former, vegetated state of the modern desert areas and other degraded land. A conservation ethic must be built on a thorough understanding of how rangeland health affects all living creatures in Iceland, from the life in the soil to the birds in the sky.

Community involvement has been increasing and is an important cornerstone in the foundation for creating a conservation ethic. A broad sector of the Icelandic society is now involved, directly or indirectly, in combating desertification and restoring land. These include rural and urban authorities, clubs, associations, and individual volunteers. Care has been taken to let participatory activities be an enjoyable experience, and awareness raising has been regarded as more important than the extent of the work accomplished. The hands-on experience in co-operative projects has fostered an understanding of the environmental problems, and created a belief in the restoration work.

It is important that leaders of society be directly involved in such projects, so that others may be stimulated to follow. Iceland has been fortunate, in that many of its leaders have had a personal interest in forestry and soil conservation. Even Iceland's Presidents have set a public example in this regard (see Preface, this volume).

6. CONCLUSIONS

The experience gained from 90 years of rangeland conservation efforts in Iceland is illustrative of failures and successes in conservation work in many parts of the world. A common thread is the inherent weakness of single issue, top-down approaches (see also Sanders, this volume; Narjisse, this volume; Onchere, this volume).

Locally led community involvement, based on a high degree of land literacy, is the foundation of the new conservation strategies that are evolving. Such involvement has proven a very powerful incentive, with a wide-ranging effect, from the grass root level to high political levels (Arnalds, 1999). Participation, which ranges from the planning to the implementation stages of projects, greatly stimulates awareness and initiative and attracts government and non-government funding.

Poor rangeland health greatly affects living conditions in many parts of the world, and Iceland is no exception. Halting soil erosion and restoring fertility of degraded land must be a national priority so that food security, economic stability and profitability may be retained. The role of soil and biota as one of the tools in meeting the Climate Change agenda is unquestionable. Carbon sequestration is an important part of the Icelandic National Climate Change Action Program for the period 1990–2000. This linkage between the Conventions of Desertification and Climate has become one of the biggest incentives for increased allocation of government funds to forestry and rangeland soil conservation programs in Iceland in recent years.

As in most parts of the world, the management of livestock grazing is a key determinant of rangeland health in Iceland. Clear guidelines for rangeland conservation must be set within effective environmental law and policy (Hannam, 1998 & this volume). Standards must be developed for guiding management of all lands within their capability. As a final resort, the law must enable penalties in cases of unsustainable land use. However, incentives are the most efficient means in reaching sustainability goals and encouraging restoration activities. Direct incentives in locally led programs combined with "enabling" incentives, such as education and training, and linking agricultural support to sustainability are among such tools. The long-range goal is the creation of a "conservation ethic", such that makes the sustainability of rangeland resources become a natural part of all land use activities.

REFERENCES

Arnalds, A. 1987. Ecosystem disturbance in Iceland. Arctic and Alpine Research 19(4), 508–513.

Arnalds, A. 1999. Incentives for soil conservation in Iceland. In: From Theory to Practice. Eds. D. Sanders, P.C. Huszar, S. Sombatpanit and T. Enters. Science Publishers Inc., USA. ISBN 1-57808-061-4, 360 pp.

Arnalds, A. and Rittenhouse, L.R. 1986. Stocking rates for northern rangelands. In: Grazing Research at Northern Latitudes. Ed. O. Gudmundsson). pp. 335–345. Plenum Press, New York, NY.

Arnalds, O. 1998. Desertification in Iceland. Desertification Contr. Bull. 32, 22–24.

Arnalds, O., Thorarinsdottir, E.F., Metusalemsson, S.M., Jonsson, A., Gretarsson, E. and Arnason, A. 1997. Soil Erosion in Iceland. Soil Conservation Service and Agricultural Research Institute, 157 pp. (In Icelandic. Available in English in 2000).

Arnalds, O., Aradottir, A.L. and Gudbergsson, G. 1999. Organic carbon sequestration by restoration of degraded areas in Iceland. In: Assessment for Soil Organic Carbon Pools. Eds. R. Lal, J. Kimble and R. Folleett. (In press).

Breckwoldt, R. 1988. The Dirt Doctors. A Jubilee History of the Soil Conservation Service of NSW. ISBN 0-7305-5845-2, 182 pp.

Campbell, A. 1994. Landcare – Communities Shaping the Land and the Future. Southwood Press Pty. Ltd., Sydney.

Gisladottir, G. and Preston-White, R. 1998. Policy changes on sheep farming in Iceland. In: Environmental Characterization and Change in South-western Iceland. Department of Physical Geography, Stockholm University, Sweden.

Hannam, I.D. 1998. Soil conservation policies in Australia: Successes and failures and requirements for ecologically sustainable policy. In: Soil and Water Conservation: Successes and Failures. Eds. T.L. Napier and S.M. Camboni. pp. 618–638. Soil and Water Conservation Society, Iowa.

Hurni, H. (Ed.) 1996. Precious earth: From Soil and Water Conservation to Sustainable Land Management. International Soil Conservation Organization (ISCO) and the Centre for Development and Environment (CDE), Berne, 89 pp.

Molleur, R.T. and Loser, J.R. 1998. Environmental Quality Incentives Program (EQIP): A new approach and tool to conservation and environment protection for America's farmers and ranchers. In: Proc. Soil and Water Conservation Society Annual Conference, July 1998, San Diego, CA.

Roberts, B. 1986. Who will speak for the land – Land Ethics, a necessary addition to Australian values. Earth Garden, May, 43–56.

Roberts, R. 1989. Land Conservation in Australia. A 200 Year Stocktake. Soil and Water Conservation Association of Australia, 32 pp.

Sanders, D.W. 1992. Soil conservation: strategies and policies. In: Soil Conservation for Survival. Eds. Kebede Tato and Hans Hurni. Soil and Water Conservation Society and World Association of Soil and Water Conservation. ISBN 91-067261.

Thorsteinsson, I. 1986. The effect of grazing on stability and development of northern rangelands: A case study of Iceland. In: Grazing Research at Northern Latitudes. Ed. O. Gudmundsson. pp. 37–43. Plenum Press, New York, N.Y.

Policy and law for rangeland conservation

Ian Hannam
Department of Land and Water Conservation, 10 Valentine Avenue, Parramatta NSW 2150, Australia
Tel: 61 2 9895 7976; Fax: 61 2 9895 7939; E-mail: <ihannam@dlwc.nsw.gov.au>

ABSTRACT This chapter discusses the environmental law and policy concerning Australian rangeland. Nearly 75 percent of Australia is rangeland. The majority of the land has been leased to users under various 'pastoral' Acts. Administrative responsibility lies with State and Territory governments. The approach taken by the Commonwealth, State and Territory governments to the development of law and policy concerning rangeland has lacked sustainable land management objectives. Numerous government inquiries have not been able to achieve a uniform approach to the natural resources law and policy to combat land degradation, the effects of land clearing and habitat loss. An essential component for, and progress toward rangeland sustainability can be achieved with environmental policy and law that gives specific attention to the ecological characteristics of rangeland and their sustainable limits. This paper examines existing legislation and policy and proposes an environmental law alternative to achieve sustainable land use.

Key words: ecologically sustainable, environmental law and policy, land degradation, land management, rangeland.

1. INTRODUCTION

Seventy five percent (600 mill ha) of Australia is rangeland (ABARE, 1994; Walker and Steffen, 1993), many of which are threatened by vegetation removal and land degradation (Young et al., 1986). Seventy percent of rangeland is used by Anglo-Europeans for pastoralism. Administration is the responsibility of States[1] and the majority of the land is leased under various tenure arrangements. Approximately 150 separate pieces of legislation, and 10 national conservation strategies directly or indirectly relate to management of rangeland (Bates, 1992; Fabricius, 1994). Ten global environmental conventions and treaties impinge upon rangeland management. Despite numerous land management inquiries, the policy and law through which the Commonwealth, and State governments have approached the management of

---

[1] 'States' in this paper refers to the States of Western Australia, New South Wales, South Australia, Queensland, and the Northern Territory.

rangeland have generally lacked sustainable land use objectives (Young, 1984; Boer and Hannam, 1992). The ecologic, economic and cultural aspects of rangeland are of great interest to all of the community (Harrington et al., 1984; Love, 1997) and its future conservation depends on policies and legislation which ensure its ecological sustainability. Common definitions of rangeland fail to properly recognise their ecological fragility and limits (Australia, 1996a).

2. NATURAL ENVIRONMENT

Desertification is the degradation of land and vegetation resources in arid, semi-arid and dry sub-humid environments resulting from various factors including climatic variations and human activities (United Nations, 1994; Gretton and Salma, 1996). Walker and Steffen (1993) maintain that the arid and semi-arid rangelands of Australia are characterised by: the co-existence of grasses and trees or other woody species; a vegetation distribution controlled by soil types and landscape factors; primary productivity which is related to rainfall; fire as an important factor in controlling vegetation composition and structure; and grazing as the most extensive land use. Australian rangelands cover a variety of bioregions, many containing habitat for rare, threatened and endangered species (Australia, 1996c). Since the Anglo-European occupation in 1788 there have been significant changes to rangeland biodiversity and approximately 12 percent of arid zone mammals (11 species) have become extinct. This represents 61 percent of all mammal extinctions in Australia. There are approximately 1800 flowering plant species in the Australian rangeland. Of these, six have become threatened at the local level, but many others are now considered endangered. Woody shrubs have spread and reduced the cover of perennial grasses and herbage. In 1975 it was estimated that 13 percent of Australia's rangeland was severely degraded and 42 percent was moderately degraded (Australia, 1978). Rangeland degradation includes accelerated soil erosion, increased numbers and distribution of weeds and feral animals, reduced water quality, soil salinisation, the decline in area and changes to native plant communities and decreased biodiversity. The risk of soil erosion in the rangelands has arisen through the effects of increased grazing pressure, opportunistic cropping in marginal rainfall areas and by disturbing large areas of land for infrastructure development (Australia, 1995; Gretton and Salma, 1996).

3. PASTORAL LAND SETTLEMENT

In the past, rangeland conservation has not ranked high as a national policy priority. Policies have focused on the allocation and administration of pastoral leases through a unique system of 'pastoral land legalisation'. Pastoral land tenure has evolved through several distinct stages of land allocation and award of property rights and was initiated as an expedient social policy response to solve the physical environmental challenges of land allocation in inland Australia. This form of tenure evolved as a powerful land control instrument and early land use policies favoured land control and settlement over environmental management (Holmes and Knight, 1994). Australia has prepared a draft national strategy for rangeland management and it recognises that problems faced in the rangeland cannot be resolved without changes to environmental policy, legislation and institutional structures (Harrington et al., 1984; Australia, 1996a). Seventy percent of rangeland is managed as leasehold land and vacant crown land under pastoral land legislation. Freehold land accounts for 17 percent while about six percent is Aboriginal reserves or Aboriginal leasehold. Seven percent is held in nature conservation reserves.

Many public inquiries have been held into pastoral land tenures but their focus has generally been on the long-standing relationship between the State as 'landlord' and the 'lessee' as the tenant with little attention paid to land degradation and biodiversity conservation issues (Holmes and Knight, 1994). Rangeland pastoralism is a low intensity land use, but individual pastoral enterprises occupy very large areas of land. Pressure to excise more land for conservation and reservation purposes has increased from conservation organisations and some government departments. Overall, there has been no systematic re-appraisal of the rationale for leasehold tenure or the need for a new approach to the environmental law and policy applying to these areas. The South Australian *Pastoral Land Management and Conservation Act 1989* is an exception, and provides for monitoring the condition of pastoral land, prevention of land degradation, flora and fauna conservation, land rehabilitation, and for Aboriginal people to follow traditional pursuits on pastoral lands.

4. ENVIRONMENTAL LAW AND POLICY FOR RANGELAND

There is a need to reform the environmental law for rangeland in Australia to improve the environmental accountability of the pastoral land law regime which has prevailed for the last century, with a focus on the protection of private and individual rights, rather than rangeland conservation (Boer and Hannam, 1992). Most States have environmental law, other than the pastoral Acts, which is concerned with the public interest and which has developed some time after the tenure-related law. These could take a more prominent rôle in rangeland management. Many of these laws embody, or are based on ecologically sustainable concepts and place a duty on decision-makers to consider the environmental, social and economic consequences of their actions (Bates, 1992; Boer, 1995).

In recent years, increasing priority has been given to the preservation of valued habitats and landscapes, and rare and endangered species. This has been spelt out in national environmental policies, including *The National Strategy for Ecologically Sustainable Development* (Australia, 1993) and *The National Strategy for Conservation of Biological Diversity* (Australia, 1996b). The draft *National Strategy for Rangeland Management* recognises that the problems in rangelands cannot be easily resolved with current legislation and institutions, present knowledge and existing dispute resolution procedures. The Strategy (although revised a number of times it was still a draft in May 1999), is a vision for the rangelands and it addresses the key issues of: policy, legislation and administration; commercial use; land management; resource conservation; cultural and heritage conservation; community needs; and research, monitoring and coordinated planning (Australia, 1996a).

5. REGULATION OF LAND USE IN RANGELAND

Many of the primary pastoral Acts and other, or secondary legislation such as the soil conservation Acts, contain regulatory provisions for the environment, but they have not been effectively utilised to control the effects of grazing and land degradation in rangeland (Boer and Hannam, 1992). The New South Wales *Western Lands Act* was introduced in 1901; although the leases were subject to land management provisions, it was administered with a "leaseholder-first" philosophy. In 1989 new environmental provisions were introduced with better control over stocking rates, soil and vegetation conservation, and land degradation. The leases were granted either in perpetuity or for a period up to 40 years for grazing, agriculture or mixed farming (Farrier, 1993). Any person who contravenes the provisions of the *Western Lands Act 1901* is guilty of a criminal offence, and remedial measures can be taken at the lessees expense or a lease can be forfeited if conditions are not complied with (Farrier, 1993). The South Australian *Pastoral Land Management Act* was introduced in 1989 to '*ensure that all pastoral land is well*

managed and utilised prudently so that its renewable resources are maintained and its yield sustained. The Act has provisions to monitor the condition of pastoral land, prevent land degradation or the degradation of indigenous plants and animals. The carrying capacity of the land is assessed in accordance with recognised scientific principles and there is a general onus on lessees to apply good land management. Under this Act pastoral land cannot be made freehold. All pastoral leases are granted for up to 42 years with land management provisions for stock control, maintenance of fencing, and land rehabilitation. The Western Australian *Land Act 1933* provides for Crown land within the State which is not withdrawn from selection for pastoral leases, and which is not required to be reserved, to be leased for pastoral purposes. A pastoral board assesses the pastoral capability with an emphasis on grazing viability rather than ecologically sustainable grazing capability. If the board decides that the land is not capable of carrying an estimated number of livestock, when 'fully developed', for the lease to be worked as an economically viable unit, it is prevented from granting the lease. Under this Act, there is a responsibility on lessees to develop plans for 'improvements', being activities to increase the rate of stocking, and land clearing, without environmental controls. The *Soil and Land Conservation Act 1945* of Western Australia has powers to prohibit the grazing of livestock on all or part of a pastoral lease for land degradation reasons and the *Land Act 1993* is empowered to reduce stock numbers on pastoral leases.

6. RURAL LAND PROTECTION ACTS

This legislation exists in most States, and establishes rural land protection boards with the power to levy land rates based on carrying capacity. The rates cover costs of rural services and a number of regulatory functions important to ecosystem management and biodiversity protection, including: drought declaration; wild dog and noxious insect management; animal health services; rabbit control, and fencing. Failure by landholders to control rabbits, wild dogs and feral pigs may lead to prosecution. Feral animals which threaten indigenous fauna can be declared noxious, for control purposes (Bates, 1992). The boards manage reserves and travelling stock routes, which in many areas have remnant vegetation which is significant for the protection of biodiversity.

7. CLEARING

The most serious environmental impacts to Australian rangeland have come from clearing and disturbing the native vegetation (McTainsh and Boughton, 1993; Australia, 1996c). The policy and legislative provisions covering native vegetation management vary significantly between the States. Past policies encouraged clearing but there has been significant change in this legislation in recent years. In Queensland, the commitment to control vegetation removal under the *Land Act 1962* has been varied, and low penalties were no deterrent to indiscriminate clearing of semi-arid vegetation. Land development schemes targeted specific vegetation communities, and some e.g., *Brigalow*, *Casuarina cristata*, *Acacia anuera*, are now below ten percent of their original extent. Under the new *Land Act* of 1994 land can be declared as an 'environmentally sensitive area' for vegetation purposes, and there are guidelines to control broadscale tree clearing on the leasehold land (Dendy and Murray, 1996). The South Australian government once encouraged a policy of land clearance for arable and pastoral agriculture and perpetual leases were issued with a standard condition for clearing (*Crown Lands Act 1929*). As a result about 80 percent of the agricultural land in the rangeland has been cleared of its original native vegetation and in several regions less than five percent remains with devastating effect on flora and fauna (Department of Environment and Land Management, 1993; Dendy and Murray, 1996). A voluntary 'Heritage Agreement' scheme was introduced in the 1970's which offered financial incentives to encourage farmers to enter into a conservation agreement to manage stock and native vegetation. The *Native Vegetation Management Act*, introduced in 1985, retained a property agreement scheme, but also included provisions for compensation and acquisition. This Act was reformed in 1991 as the *Native Vegetation Act* with its broader and more environmentally cognisant land management provisions.

The most recent native vegetation Act in Australia is the New South Wales *Native Vegetation Conservation Act 1997* (Department of Land and Water Conservation, 1997a). This Act is more advanced than the South Australian Act with extensive provisions to enable the whole community to participate in native vegetation surveys, planning and decision-making, and in this context is a significant piece of rangeland management law in Australia. This Act establishes 'biogeographic regions' and a statutory regional vegetation management plan is required to be prepared for each region. The plans delineate different areas for which specific land management conditions apply, including areas of protected vegetation, areas where vegetation may be cleared subject to comprehensive environmental assessment, and areas where minor disturbance may take place without development consent.

The Act establishes the native vegetation trust fund and money is allocated to landholders who voluntarily enter into a comprehensive property agreement for vegetation conservation and protection, including sustainable grazing management.

7.1 Native grassland and shrubland

Australia's rangeland grassland communities have been extensively modified and degraded (McTainsh and Boughton, 1993). In an effort to prevent further loss, the definition of native vegetation in the New South Wales Act includes 'groundcover', which is vegetation dominated by herbaceous species if not less than 50 percent of the herbaceous vegetation in the area comprises indigenous species (usually from the families *Poaceae*, *Asteraceae*, *Cyperaceae*, *Chenopodiaceae* and *Fabaceae*, Department of Land and Water Conservation, 1997b). Recent prosecutions under this Act for illegal clearing in rangeland have now created new legal and ecological standards for native shrubland-grassland community conservation.

8. SOIL CONSERVATION LEGISLATION

Some Australian soil conservation Acts contain provisions for ecosystem protection, but they have only been applied to a limited extent for rangeland conservation (Bradsen, 1988). Soil conservation legislation has been slow to address these vegetation issues. Because the remedies require direct interference with traditional property rights and land management practices, which is difficult politically (Bates, 1992; Hannam, 1999), this has been the prompt for some State governments to introduce new forms of legislation. One example, the *New South Wales Native Vegetation Conservation Act*, emphasizes incentives, community conservation projects and improved education, rather than regulation. A few of the States have linked soil and native vegetation conservation legislation, a logical approach to manage land degradation (e.g., South Australian *Soil Conservation and Land Care Act 1989* and the *Native Vegetation Act 1991*). The New South Wales *Soil Conservation Act 1938* has a unique provision to declare areas of 'erosion hazard' which can be used to prevent over-grazing, and wind or water erosion in semi-arid ecosystems.

9. RANGELAND CONSERVATION REFORM

It is appropriate that the Commonwealth government develop a national model for rangeland conservation legislation, for uniform adoption across Australia. The precedent for this approach has been established for other areas of environmental law e.g., threatened species. Such an approach would see the ecological principles advocated in global environmental strategies being adopted in the model (Boer, 1995; Sands, 1995). A key objective of these strategies is that ecologically sustainable principles could be used to influence the legal concepts, definitions and the operational provisions of State rangeland policy and legislation (Australia, 1994). Principles from the following global strategies could be used:

- *The Rio Declaration on Environment and Development* (United Nations, 1992a).
- *Agenda 21* (United Nations, 1992b).
- *The Convention on the Conservation of Biological Diversity* (United Nations Environment Program, 1992).
- *The Convention to Combat Desertification* (United Nations, 1994).
- *The Draft International Covenant on Environment and Development* (IUCN Commission on Environmental Law, 1995).

These strategies include important environmental elements which apply to rangeland management reform and include provisions: for community participation in environmental planning; to ensure that social, cultural, heritage and biodiversity characteristics are assessed; and to improve the accountability of rangeland management institutions to the community and the government. The global argument for individual nations to make a serious commitment to the formulation of national sustainable policies and strategies and to support these with appropriate structures (Carew-Reid et al., 1994; Sands, 1995) is already reflected in the following Australian documents:

- *The Intergovernmental Agreement on the Environment* (Australia, 1992a).
- *The National Strategy for Ecologically Sustainable Development* (Australia, 1993).
- *The Draft National Strategy for Rangeland Management* (Australia, 1996a).

These national environmental strategies have been generally well-received by the States. However, with the current extent and severity of land degradation in rangelands, a fundamental change in attitude in society toward rangeland conservation is essential before major changes can be expected in their condition (Ludwig and Freudenberger, 1997). Many countries have developed a capacity to implement sustainable development. In Australia, its actual achievement will not only be dependent on the political will of the Commonwealth and State governments to develop adequate legislation, but to put it into practice through increased incentives for property management and better technical support from government institutions (see also Sanders, this volume). Reforming rangeland conservation law in this manner can give special consideration to biodiversity conservation and the sustainable land use objective (Hannam, 1992). Any legislation and policy should be based on the major ecological issues facing Australian rangeland, including vegetation management, land degradation and maintenance of biological diversity and it should be able to resolve multiple use conflicts.

Broad strategies and objectives accompany the nine key goals of the National rangeland strategy and these could form the foundation of a new policy and legislative regime. Legislative provisions can be developed to: distinguish the responsibilities of administrators and all rangeland users; establish resource assessment methods and indicators for ecologically sustainable rangeland management; provide flexibility in land use; monitor condition and degradation; develop bioregional conservation strategies to protect biological diversity; recognise and protect the rights of indigenous peoples; recognise social, cultural and heritage aspects; research environmental, economic and social aspects of rangelands; and develop land use planning guidelines (Australia, 1996a). Rangeland management legislation must be supported by community education programs such as the National Landcare Program which is the main land management support group in Australia (Alexander, 1995). Working in partnership, government officers, landholders and citizens can develop vital information to improve land management in rangelands. Education material should include: guidelines to assess the value of ecologically sustainable standards/indicators; a review of the ability of land management technology to reach and maintain ecologically sustainable standards; and indicators to detect where ecologically sustainable land use practices are being successfully applied (Bradsen, 1994; Carew-Reid et al., 1994; A. Arnalds, this volume).

10. LEGISLATIVE MODEL

Reform of rangeland legislation and policy should be consistent with any general reform of land management legislation in Australia (Gardner, 1994). Three approaches which could be considered by the Commonwealth and put to the States include: broad-based integrated land management legislation; specialised rangeland legislation (operating in conjunction with the pastoral Acts); or various combinations of these. The aim of integrated legislation is to manage a number of natural resources, generally including water, soil, national parks, native vegetation and threatened species, under the one Act.

Its main advantage is that it integrates all aspects of ecosystem management in a single assessment and decision-making process, including land planning, incentives, licensing, property planning, education, research and enforcement. The Queensland *Nature Conservation Act 1992* is an example of integrated legislation. A disadvantage of this type of legislation is that it requires a 'mega department' bureaucracy to administer which increases the propensity for competing interests to intervene in land management decisions. Specialised 'rangeland management' legislation would include the key components necessary for rangeland ecosystem management, including provisions for rangeland research and education, rangeland assessment and planning, protected area management, and biodiversity protection. Current legislative models which are based on ecosystem conservation include the *Native Vegetation Act 1991* and *Pastoral Land Management Act 1989* of South Australia. This approach is favoured over the 'integrated approach' because the legislation would focus on the rangeland and such an Act would link with a 'regime' of supportive environmental law for the rangeland environment, including national parks and wildlife, environmental planning and assessment, threatened species conservation, natural heritage and soil and water conservation. This legislation would be administered by a land management agency with specialised rangeland land use expertise.

11. KEY PROVISIONS

There are a number of key provisions which States can apply to achieve a general objective of ecologically sustainable rangeland management. These include legislative provisions which: establish a general duty of care ethic, a right to take jurisdiction over rangeland, establish rules and criteria for land management, establish mechanisms to create cooperation in rangeland management, decide ultimate or primary responsibility for various groups, and determine the circumstances for intervention. The objects of sustainable land management legislation must be underpinned by ecosystem-based definitions, similar to those in the New South Wales *Native Vegetation Conservation Act 1997*, and cover at least: rangeland; rangeland conservation; biological diversity; and ecologically sustainable use. Public opinion would be invited through provisions which: constitute a community-based rangeland advisory body; require the preparation of a State rangeland strategy; and require public exhibition of rangeland management plans. Incentives to encourage voluntary preparation of property agreements which contain ecologically sustainable techniques are a key component of the legislation (Australia, 1996a).

11.1 Bioregional plans

Regional plans and strategies can be prepared for one, or a number of specific rangeland ecosystems. The plans set out in detail, administrative, policy, ecological, land capability, land use and sociological information (Farrier, 1993). The objective is to derive a statutory plan, which divides the region into a series of land units according to their ecological capabilities. Land management policies, guidelines and acceptable standards of land use may be developed for each land unit, including the circumstances where an environmental impact assessment may be required (Bosch and Booysen, 1992). A regional plan may specify when particular land use activities (e.g., use of fire) are prohibited altogether, or prohibited beyond a specified level.

11.2 Land management codes

These represent a self-regulation system which defines the ecological capabilities of land use, in the form of specific land management conditions and guidelines, and is applicable for on-going routine land use. Although under-utilised to date in Australian environmental law, this mechanism can be more widely applied for rangeland land use. Codes also denote land use circumstances where an approval, or a license based on an environmental assessment, may be required.

11.3 Property agreements

This is a voluntary system where a landholder and a local authority agree, through a formal instrument, to a prescribed course of management over a specified time period. As a 'two-way' system, if the landholder agrees to specific conservation and land management practices, including cessation of agricultural use in some areas, the landholder could be eligible for financial and technical support (e.g., training for land managers in land assessment and biodiversity surveys). Property agreements offer one of the best mechanisms for improving rangeland management as they bring parties together with a common commitment and a common community goal to reduce land degradation and to protect biodiversity.

11.4 Environmental planning and assessment

The legislation may include an approval system for activities with a potentially high environmental impact, involving a major change or an intensification in land use, e.g., clearing extensive areas of native vegetation, or changing the direction of surface water flow. These requirements would be defined in a bioregional plan or under a separate development control system (Farrier, 1993), such as used in the New South Wales *Native Vegetation Conservation Act 1997*, where, depending on circumstances, an environment assessment, an environmental impact statement or threatened species impact statement (under Threatened Species legislation), may be required to be prepared. Most States have adequate environmental planning and assessment legislation and some (e.g., New South Wales *Environmental Planning and Assessment Act 1979*), contain environmental assessment procedures suitable to address rangeland conservation issues.

11.5 Conservation areas and ecological reserves

Conservation easements provide a flexible means of achieving biodiversity protection from lessees or landholders (Bates, 1992). Regional analysis will pinpoint significant biodiversity conservation areas, and holdings with high conservation values can be targeted. If the area is leasehold, conservation conditions can be added to the lease. General ecological criteria for selecting conservation reserves can be established in a bioregional plan. The aim of establishing conservation areas and ecological reserves is to protect an adequate representation of each rangeland ecological community (West, 1993). Availability of incentives, such as rate relief and financial and land management support under a property agreement, will increase the attractiveness of elective conservation.

12. CONCLUSION

Significant ecological improvement of Australia's rangeland by reducing land degradation and improving biodiversity protection can be assisted by more effective environmental law and policy and a better land management system. The Commonwealth Government should take the initiative and produce model legislation and policy guidelines for the States to adopt. The plethora of national and State conservation strategies, combined with key global conservation strategies, provide excellent source material to develop the strategic rangeland management policy. There is an adequate amount of existing environmental law (i.e., threatened species, native vegetation, soil conservation, environmental planning and assessment, land tenure law) with suitable types of ecological provisions to draw upon to help frame the environmental law regime for better rangeland management. The approach canvassed here should be considered for Australia and it may offer some guidance to other nations.

REFERENCES

ABARE (Australian Bureau of Agricultural and Resource Economics) 1994. Rangelands Report.

Alexander, H. 1995. A Framework for Change, the State of the Community Landcare Movement in Australia. A National Landcare Facilitator Project. Annual Report. National Landcare Program.

Australia 1978. A Basis for Soil Conservation Policy in Australia. Commonwealth and State Government Collaborative Soil Conservation Study. 1975–1977. Department of Environment, Housing and Community Development. Australian Government Publishing Service, Canberra.

Australia 1992a. Intergovernmental Agreement on the Environment. Heads of Government in Australia.

Australia 1993. The National Strategy for Ecologically Sustainable Development. Australian Government Publishing Service, Canberra.

Australia 1994. Australian Initiatives to Combat Desertification. Department of the Environment, Sport and Territories, Canberra.

Australia 1995. Native Vegetation Clearance, Habitat Loss and Biodiversity Decline. An Overview of Recent Native Vegetation Clearance in Australia and its Implications, Biodiversity Series. Paper No. 6. Biodiversity Unit. Department of the Environment, Sport and Territories.

Australia 1996a. Draft National Strategy for Rangeland Management. Australian and New Zealand Environment and Conservation Council and Agriculture and Resource Management Council of Australia and New Zealand.

Australia 1996b. The National Strategy for the Conservation of Australia's Biological Diversity. Department of Environment, Sport and Territories. Australian Government Publishing Service, Canberra.

Australia 1996c. State of the Environment. Australia. An Independent Report Presented to the Commonwealth Minister for the Environment by the State of the Environment Advisory Council. Department of the Environment, Sports and Territories.

Bates, G.M. 1992. Environmental Law in Australia. Butterworths, Sydney.

Boer, B.W. and Hannam, I.D., 1992. Agrarian land law in Australia. In: Agrarian Land Law in the Western World. Eds. W. Brussaard and M. Grossman. pp. 212–233. C.A.B Wallingford, Oxon.

Boer, B.W. 1995. Institutionalising ecologically sustainable development: the roles of national, state, and local governments in translating grand strategy into action. Willamette Law Review 31, 2307–2358.

Bosch, O.J. and Booysen 1992. An integrative approach to rangeland condition and capability assessment. Journal of Rangeland Management 45(2), 116–122.

Bradsen, J.R. 1988. Soil Conservation Legislation in Australia. Report to the National Soil Conservation Program. University of Adelaide.

Bradsen, J.R. 1994. Natural resource conservation in Australia: some fundamental issues. In: Adopting Conservation on the Farm, An International Perspective on the Socioeconomics of Soil and Water Conservation. Eds. T.L. Napier, S.M. Camboni and A. El-Swaify Samir. pp. 435–460. Soil and Water Conservation Society, Ankeny, Iowa.

Carew-Reid, J., Prescott-Allen, R., Bass, S. and Dalal-Clayton, B. 1994. Strategies for National Sustainable Development, A Handbook for their Planning and Implementation, IUCN, The World Conservation Union, Gland Switzerland.

Commission on Environmental Law 1995. Draft International Covenant on Environment and Development. Environmental Policy and Law Paper No. 31. IUCN Environmental Law Centre. IUCN, Gland Switzerland.

Dendy, T. and Murray, J. (Eds.) 1996. From Conflict to Conservation. Native Vegetation Management in Australia. A Focus on the South Australian Program and Other Australian Initiatives. South Australian Department of Environment and Natural Resources.

Department of Environment and Land Management 1993. State of the Environment Report. South Australia.

Department of Land Water Conservation 1997a. A Proposed Model for Native Vegetation Conservation in New South Wales.

Department of Land and Water Conservation 1997b. Definitions and Exemptions. State Environmental Planning Policy No. 46. Amendment No. 2.

Fabricius, C.P. 1994. Guide to Environmental Legislation in Australia and New Zealand. A Summary and Brief Description of Environment Protection and Related Legislation of the Federal, State and Territory Governments of Australia and New Zealand. Report No. 29. Australia and New Zealand Environment and Conservation Council.

Farrier, D. 1993. The Environmental Law Handbook. Planning Land Use in New South Wales. Redfern Legal Centre Publishing, New South Wales.

Gardner, A. 1994. Developing norms of land management in Australia. The Australasian Journal of Natural Resources Law and Policy 1(1), 149–157.

Gretton, P. and Salma, U. 1996. Land Degradation and the Australian Agricultural Industry. Industry Commission. Australian Government Publishing Service.

Hannam, I.D. 1992. The concept of sustainable land management and soil conservation law and policy in Australia. In: Proceedings of International Conference on Sustainable Land Management. Ed. P. Henriques. pp. 153–168. International Pacific College, Palmerston North, New Zealand.

Hannam, I.D. 1999. Soil conservation policies in Australia: successes and failures and requirements for ecologically sustainable policy. In: Soil and Water Conservation Policies: Successes and Failures. Eds. T.L Napier and S.M. Camboni. Soil and Water Conservation Society, 7515 Northeast Ankeny Road, Ankeny, Iowa.

Harrington, G.N., Wilson, A.D. and Young, M.D. 1984. Management of Australia's Rangelands. CSIRO, Melbourne, Australia.

Holmes, J.H. and Knight, L.R. 1994. Pastoral lease tenure in Australia: historical relic or useful contemporary tool? The Rangeland Journal 16(1): 106–121.

Love, M. 1997. The farmgate effect. In: The Wik Case. Issues and Implications. Ed. G. Hiley. Butterworths, Sydney.

Ludwig, J. and Freudenberger, D. 1997. Towards a sustainable future for rangelands. In: Landscape Ecology. Function and Management. Principles from Australia's Rangelands. Eds. J. Ludwig, D. Tongway, D. Freudenberger, J. Noble and K. Hodgkinson. pp. 121–131. CSIRO, Australia.

McTainsh, G.H. and Boughton, W.C. (Eds.) 1993. Land Degradation Processes in Australia. Longman, Cheshire, Melbourne.

Sands, P. 1995. Principles of International Environmental Law. Volume I. Frameworks, Standards and Implementation. Manchester University Press, Manchester.

United Nations Environment Program 1992. Convention an Biological Diversity, Strategy for Biodiversity Conservation.

United Nations 1992. Earth Summit. United Nations Conference on Environment and Development. Rio de Janeiro, Brazil, Regency Press, London.

United Nations 1992a. The Rio Declaration. United Nations, New York.

United Nations 1992b. Agenda 21. United Nations, New York.

United Nations 1994. Convention to Combat Desertification. UNEP, Nairobi, Kenya.
Walker, B.H. and Steffen, W.L. 1993. Rangelands and global change. The Rangeland Journal 15(1), 95–103.
West, N.E. 1993. Biodiversity of rangelands. Journal of Rangeland Management 46(1), 2–13.
Young, M.D. 1984. Rangeland administration. In: Management of Australia's Rangelands. Eds. G.N Harrington, A.D. Wilson and M.D. Young. pp. 157–170. CSIRO, Melbourne, Australia.
Young, M.D., Walker, P.A. and Cocks, K.D. 1986. Land use in Australia's rangelands. Australian Rangeland Journal 8(2), 131–139.

Rangelands issues and trends in developing countries

Hamid Narjisse
Institute of Agronomy and Veterinary Sciences, Hassan II, B.P 6202 Rabat, Morocco
Tel: 212 7 77 70 18; Fax: 212 7 68 03 97; E-mail: <narjisse@mtds.com>

ABSTRACT Although rangelands in developing countries are of low productive potential, they remain of vital importance for the survival of a growing population. These lands are degraded as a result of human mismanagement, heavy pressure of various origins and conflicting uses. Efforts to mitigate rangeland deterioration in developing countries have failed in general. The approaches were inadequate, the policies were incoherent and lacking long-term vision, and the technical packages were either wrong or ill adapted.

Prospects for rangeland recovery and sustainability in developing countries are uncertain. Demographic pressures and lack of stewardship are among the driving forces fuelling this uncertainty. Fortunately, we currently witness the emergence of promising ideas and practices in some developing countries. These include improved democracy and governance, bottom-up approaches and integration of multi sectoral actions. The progressive adoption of these concepts will contribute immensely to the successful transition to sustainable rangeland rehabilitation and development.

Key words: degradation, developing countries, holistic, land use, policies, sustainability.

1. INTRODUCTION

Rangelands are usually marginal lands located in arid ecosystems. Nearly 30% of the world's land surface is semi-arid to arid rangeland, where management units are large commercial ranches, or areas managed by nomadic, transhumant or sedentary pastoralists. Rangelands are characterized by relatively low rainfall, often unreliable and variable from year to year. As a consequence of low levels of output, rangelands are often economically and politically marginal, with poorly developed or limited physical and economic infrastructure. The chronic poverty prevailing in these areas acts as an aggravating factor.

The most direct effect of desertification is the decline in the quality and quantity of natural assets: soil, water and biodiversity. The value of the production lost annually in developing countries (DCs) has been estimated by UNEP to be $16,000 million (Nessim, 1996). This estimate does not include the off-site costs of desertification, such as downstream siltation of dams, which are reported to be of high magnitude, costing for example as much as $(US) 60 millions per year in Morocco (Narjisse, 1996a).

Development of the range/livestock sector in DCs, for a long time followed the classical paradigm promoting sedentarization, privatization and intensification. Most experts agree that this option has proven impractical and incompatible with the reality of these countries. This stimulated the emergence of a new paradigm based on participatory approach to planning and managing common use of natural resources, promotion of local knowledge and integration of rangeland development into the broader framework of rural development (see also A. Arnalds, this volume).

The purpose of this chapter is to present a comprehensive review of the issues and trends in rangeland condition and use in DCs, especially those located in North Africa and the Sahel. However, many of the ideas discussed may be applicable to arid and semi-arid rangelands in other DCs as well.

2. CAUSE AND EXTENT OF RANGELAND DEGRADATION IN DCs

The extent and severity of land degradation is poorly documented and there is a considerable controversy over the rate and extent of land degradation actually occurring in drylands. In fact, Biot et al. (1992) indicated that soil loss might have been overestimated by a factor of ten. If inaccurate estimates are used as the basis for policy formulation and planning, the potential for differences in perception and conflict between policy-makers and land users is obvious.

Until recently, overgrazing associated with traditional pastoralism and common property land tenure systems was considered the prime cause of rangeland degradation. This is an oversimplification of a complex problem. The spread of poverty, the breakdown of common property rules and the abandonment of herd mobility practices are certainly important driving forces of rangeland deterioration. It is therefore more appropriate to consider rangeland desertification as a cumulative outcome of unsustainable land use practices, which are the result of incoherent national policies (see also Sanders, this volume). These, in turn, may be exacerbated by impacts of international policies and natural disasters.

2.1 Unsustainable land use practices

Unsustainable land uses include expansion of rainfed cultivation onto unsuitable lands, soil and ground water mining, overgrazing, and uncontrolled harvest of plant biomass. The adoption of these practices by those who depend on dryland resources for their livelihood is usually a rational and optimal response to prevailing incentives and constraints.

2.1.1 Overgrazing

Overgrazing is a common practice in most DCs. Authors from Morocco, Algeria, Tunisia and Sahelian countries have reported stocking rates three to five times higher than estimated optimum carrying capacity (MAMVA, 1995; Aidoud, 1994; Kallala, 1994; Soumare, 1996). Analysis of the production systems in these countries suggests that overgrazing is primarily caused by a few large producers of the community who possess the means to take advantage of the grazing community. Another important factor contributing to overstocking results from the lack of credit possibilities, which forces pastoralists to retain as many animals as possible to secure renewal of their only source of income, the herd, after the occurrence of natural disasters.

2.1.2 Agricultural expansion

Population demands have increased the pressure on existing productive land and have forced crop production onto increasingly marginal land. This process is accompanied by an increasing rate of soil and vegetation degradation. An assessment of land use change occurring in Tunisia from 1880 to 1980 indicates that cultivated land increased by three million hectares, while the area used as grazing land declined by four million hectares (UNEP, 1996). Similarly, Bilsborrow and Okoth-Ogendo (1992) reported that land extensification in Latin America also represented an important contributor to agricultural output increase (66%) between 1950 and 1975.

Cropland expansion is generally encouraged by incoherent pricing and subsidy policies and is exacerbated by the land tenure regime. Land cropping is usually the first step toward common land's ownership. Customary law in many Islamic countries grants ownership of land to a person who has exploited the land for a given period of time without being challenged for the illegality of such use.

2.1.3 Fuelwood harvest

Concerns are often expressed in dryland areas over a reduction of tree and shrub cover and hence an impending firewood shortage. For example, in the Machakos region of Kenya, reports in the 1940s, 1950s and 1970s all estimated that the usage of wood was greater than the annual increment (Mortimore, 1992). Similarly, throughout Southern and Eastern Mediterranean countries, harvest of firewood for domestic and sometimes commercial uses is a major cause of desertification. In Morocco, for example, the average household consumption of firewood is estimated at approximately 2.61 tons per year (MAMVA, 1994). Such levels of biomass harvest exceed the production potential of most Moroccan steppe and woodland ecosystems. FAO (cited by Naveh, 1994) projects that by the year 2000, Southern and Eastern Mediterranean countries will suffer a deficit of about 30 million m^3 of wood to supply their needs for cooking and heating. Taking into consideration that wood production in these countries' woodlands range from 0.5 to 1 m^3 ha^{-1} $year^{-1}$, this may potentially endanger all the remaining woodland of this region. For example, approximately 25,000 hectares of Moroccan woodlands are lost each year to compensate for the excessive fuelwood harvest (Narjisse, 1996a).

2.1.4 Urbanization

Developing countries affected by desertification not only face high population growth rates but also high rates of urbanization (8–10% per annum). Some countries in Latin America already have 3/4 of their population living in towns and cities, with Asia and Africa just above 1/3 and under 1/3, respectively (Nessim, 1996).

Such developments place a tremendous burden on governments, which have to provide costly social services at the expense of productive infrastructures. They also lead to a heavy strain on urban surroundings, especially in terms of depletion of prime agricultural lands. The corollary of uncontrolled urbanization is poor management of land and other natural resources, resulting in more land degradation.

2.2 Demographic processes

It is clear that rapid population growth in DCs leads to increased pressure on the land and other resources. In the Sudano-Sahelian region, population increased by almost 25% between 1977 and 1984, while in North Africa the annual increase was on average higher than 2% with a decreasing tendency (Soumare, 1996). In most of these countries, population growth rates are usually higher than the most optimistic predictions for economic growth, and hence deepen the poverty conditions.

Population growth, poverty and environmental degradation feed each other and are all the ultimate manifestation of underdevelopment. Thus, environmental degradation may increase the incentive to have larger families, as more household time is required to harvest common property goods, and expanding cultivation onto marginal lands is the only way to increase food production to meet increasing household needs.

2.3 Institutional processes

Institutional processes are affected by national and international driving forces, and the various biases such as gender or urban. Among these processes, the following are considered to be of particular significance to rangelands.

2.3.1 Tenurial ambiguities

The nature of tenurial arrangements is a prominent factor in the sustainability of land use patterns. Most rangelands in DCs have a common tenure regime, which is considered to preclude investment to improve these lands. The "tragedy of the commons" argument was very popular within governments and development circles for a long time, but recent research has documented the limitations of this argument (Niamir-Fuller, 1996). The poverty/environmental degradation link is accentuated by inequitable distribution of land, and by ambiguities and uncertainties of communal tenure rights.

The terms upon which access to common property resources is gained determines the particular group of producers whom it will benefit. In general, because the traditional management of common property resources is often extremely hierarchical, the poorest groups are not offered the opportunity to participate in the use of commons. In many cases, households may be excluded from gaining access to these resources because they do not have the means to do so. For example, the costs associated with herd acquisition and/or livestock watering are high enough to significantly restrict the access to many.

2.3.2 Infrastructure and government support services

Provision of roads and other communication infrastructure is usually poor in rangeland areas. This results in limited access to various inputs. Lack of infrastructure also inhibits the development of markets, which can be considered a *sine qua non* condition of development. Malfunctioning markets contribute significantly to rangeland desertification (Olsson, 1993), by restricting movements of goods and limiting options for income generation from off-farm enterprises.

Lack of adequate health services, basic education, water and sanitation, also affects the productivity and ultimately the income of rural households, increasing their vulnerability to recurrent droughts. Absence of electricity and water supplies may also act as an incentive for maintaining large families that are required to meet household water and energy needs.

2.3.3 Lack of adequate local organizations

Rangeland areas have often experienced a decline of traditional organizations' authority, aggravated by the deterioration of local government institutions. The weakening of internal community institutions and mechanisms seriously limit the ability to control members' access to commonly owned resources. After independence, the new governments promoted centralisation of political powers, sometimes deliberate abolishment of local institutions, and sedentarization of mobile populations. This resulted in reduced mobility of herds and led to overgrazing of land surrounding the settlements.

Many studies emphasize the merits of refining and adjusting, rather than eliminating, traditional pastoral range management practices (Soumare, 1996). Strong and effective local institutions are arguably a major prerequisite for sustainable rangeland development (see also A. Arnalds, this volume). The absence of operational grassroots organizations is an additional constraint to effective management of community resources and to successful resolution of conflict in their use. The lack of such organizations limits the prospect for participatory development and the devolution of decision-making to the community level.

2.4 National and international policies

In order to address macro-economic distortions and redress budget deficits, many DCs adopted structural adjustment programs. The effects of such programs on rural livelihood are country-specific. In general, liberalization policies have adversely affected the short-term ability of governments to invest in the socio-economic development of dryland areas, reinforcing existing urban biases.

2.4.1 Pricing

Some public policies, such as direct financial support to grain growers, promote the conversion of rangeland to cultivated use. Similarly, direct and indirect controls of meat prices are important constraints. For example, importing of low cost beef and mutton into Africa constitutes unfair competition to the local producers, while protection of local meat production in other DCs increases the pressure on rangeland resources (Narjisse, 1996b). Application of market prices to outputs and inputs is a key condition to sustainable rangeland development.

Pricing policies may also favor drought sensitive crops. In many DCs, market distortion policies led to the replacement of drought-tolerant crops such as barley, with wheat in areas which were not agro-ecologically suited for wheat production. Such practices have always increased dependency on resource mining to compensate for production shortfalls.

2.4.2 Subsidies

Inappropriate sectoral agricultural policies may fail to take into account critical intersectoral, intrasectoral and inter-regional linkages. These include pricing policies favoring expansion of cereal production into marginal areas, subsidized capital to expand uneconomic commercial operations, and subsidies for the adoption of inappropriate technologies (Narjisse, 1996b).

In North Africa and the Middle East, governments heavily subsidized concentrate feed, especially during drought. There is a consensus that such policies foster range degradation by enabling stocking rates to be maintained at artificially high levels. More obvious is the effect of subsidies on mechanization, which has led to the plowing and severe degradation of marginal areas as observed in the steppic regions of Morocco (Narjisse, 1996a) and Algeria (Aidoud, 1994).

2.4.3 Credit

In general, pastoralists have limited access to financial services. This is partly due to the lack of institutional development in marginal areas, and partly due to the reluctance of conventional credit agencies to lend to people, who by definition have few assets for collateral. The lack of credit accentuates the short-term planning horizons of the pastoralists caused by other factors such as the lack of secure tenure. This renders activities which reduce rangeland degradation but which also have a relatively long-term pay off, economically unattractive. The absence of saving opportunities is also a constraint. It can lead, for example, to over-capitalization in the form of livestock, and hence to overgrazing of rangeland resources.

2.4.4 Declining Commodity Prices

Agricultural commodity prices have been declining in real terms, and this trend is likely to continue. Between 1980 and 1988, the real prices of non-fuel commodities produced by developing countries declined by 40% (Nessim, 1996). The reasons for this decline are complex. They involve protectionism in developed countries, lack of diversification, biotechnological development allowing industrial production of substitute products, and increased production in some countries due to technological innovation.

In order to maintain incomes in the face of falling prices for commodities, many developing countries are increasing production. About 80% of the growth in African agriculture has come from the extension of the area under cultivation, predominately on lands once used for grazing (Kates et al., 1993). Land under export crops is expanded, resulting in soil mining, encroachment into marginal grazing land, and misuse of natural resources.

2.5 Natural Disasters

Natural disasters have been increasing in frequency and intensity over the past decades. It is estimated that 23 million African people were affected by drought in 1990/1991 (Soumare, 1996). The cost of drought relief in Africa alone was in the range of five to eight billion US dollars in the last decade (Nessim, 1996). In most arid and semi-arid rangelands, drought is a recurrent hazard that needs to be incorporated into normal planning.

Drought episodes affect primarily the rural poor and the natural resource base on which they depend. They amplify the scarcity of rangeland forage resources and hence accentuate the pressure on them. They reduce employment opportunities and consequently aggravate the poverty conditions, leading to an increase in the demand exerted on already abused and scarce natural resources. Finally, drought events favor change from pastoral way of life to cultivation among poor pastoralists, due to loss of all or most of their herds.

Besides drought, other natural disasters are common and can destabilize the livelihood systems of rural people or directly lead to desertification. The recent infestation of screwworm in Libya and the recurrent epidemics of desert locust attacks throughout dryland areas of the African continent are notable examples.

3. RANGELAND'S CURRENT CHALLENGES AND ISSUES

The challenges facing rangeland development in DCs are diverse. Some are inherent in the physical and climatic features of DCs. Others relate to institutional and policy related processes in these countries or at the international level. Managers and decision-makers have little control over most of the issues involved. The latter's understanding and prioritization is essential to reduce the uncertainty surrounding the planning of rangeland development and management.

3.1 Economic viability of pastoralism

Economic viability of pastoralism is an issue in DCs, as well many developed countries (Chisholm, 1992). Except for a few producers that managed to adjust the size of their operation, the majority of pastoralists are operating at very low levels of return. Pastoralism is indeed losing ground when compared to other uses of arid land. For example, Holdgate (1994) reported that ecotourism in Kenya, Tanzania, Zimbabwe and Costa Rica generates 40 times higher return per hectare than what can be obtained from converting the land into cattle pasture. The long-term economic viability will therefore depend on the way society views this sector. In particular, there is a need for promoting diversification including non-consumptive uses of rangeland resources and looking at these resources not exclusively from a commodity perspective. In addition, pastoralists should be compensated for undertaking appropriate activities maximizing water harvest and securing biodiversity conservation. At present, urban markets gain the benefits of such activities, while the local rural communities bear the costs.

3.2 Rangelands access

An important component of the solution to problems of rangeland degradation is our capacity to design, through a process of consultation, an institutional framework that will secure access regulation, and provide the foundation for granting the decision-making process authority to local communities assisted by adequately funded and staffed government agencies.

The issue here is who can access common rangeland and how this access should be managed. In other words, there is a need to define whether the organization of access should be inclusive to all having rights, or whether it should exclude certain members of the communities such as absentees, descents of right-holders, and those holding rights acquired through marriage. Clearly, some form of exclusiveness is likely to be necessary if any form of ecologically sound management is to be applied. Application of some of these concepts will require modifications of customary laws and rules of inheritance.

In addition to the exclusiveness issue, there is also a need for charging a symbolic but incremental grazing fee to access collective rangeland. At least three benefits can be derived from this arrangement. First, it would encourage the culling of unproductive livestock and hence reduce the pressure on rangeland resources. Second, it would introduce some equity between rangeland users. Thirdly, it would generate funds that could be used for various improvement investments. A key to the success of such a measure is to secure that payments are made to the pastoralists' community, not to the treasury. In this respect, the project on Rangeland Development in the High Plateaus of Morocco, sponsored by the International Fund for Agricultural Development, can be listed as a success story. Pastoralists from this region, organized in co-operatives, are currently collecting grazing fees charged to their members and used later to subsidize emergency feed acquisition and transport.

The corollary of introducing a grazing fee is the establishment of an operational organization that regulates access to rangeland, and negotiates matters related to rangeland development with government agencies or any other parties, on behalf of its members. Another task to be assigned to local organizations is to secure a fair and accepted arbitration between conflicting uses and interests. Achieving this objective is a real challenge, as central government control of land use can only be handed to the producers, when government at all levels recognizes that sustainable development has to be planned and implemented at the local community level.

3.3 National policies

Since the Rio conference (UNCED, 1992) and the adoption of the concept of sustainable development, new philosophy emerged, promoting ideas of local development and a balanced approach to national development, whereby conservation concerns receive more attention than in the past. In particular, it is recognized that marginal areas such as rangelands play an important ecological role and should therefore be dealt with not only from an economic perspective which considers only commodities, but also as a type of land supporting multiple uses and services to society. Some of these uses may not have readily identifiable or traditional commodity values.

The new approach to rangeland development considers this issue as a national solidarity issue, which means that more resources are allocated to natural resources conservation. In addition, rangeland development is no longer dealt with as an isolated problem. Instead, the link to poverty and the need for a comprehensive rural development framework are now widely recognized.

4. POSSIBLE TRENDS OF RANGELANDS STATUS IN DCs

Pastoral economies across the developing world are experiencing rapid change. Assessment and detection of these trends are the key to sound planning of rangeland development and land use. This is not an easy task. It requires a continuous monitoring of rangeland social, economic, institutional and biophysical attributes.

4.1 Demographic pressure, migration and urbanization

Population growth and urbanization usually augment the pressure on prime rangelands and forests, which are then converted to cropping or urban lands. However, when population growth reaches a limit, the pauperization of the landless accentuates economic incentives for rural to urban migration. Under extreme circumstances, depopulation may lead to land abandonment as was noticed in many European countries at the beginning of the 20th century (Hubert, 1991) and as is currently observed in some DCs (MAMVA, 1995).

4.2 Privatization of common property resources

Following the wave of the "tragedy of the commons" concept, privatization was promoted as the solution to rangeland problems. This resulted in the conversion of large areas into commercial ranching in Africa, Central Asia and Latin America. Currently, the privatization process has gone about as far as it can go. Rangelands with high potential have already been privatized, while the remaining area is so small in size that the ratio of land to capita does not allow any further partitioning of the land. Fragmentation of grazing units into a large numbers of small allotments is ecologically and economically inappropriate, and the alternative of few large ranches is socially unjust. Historically, privatization has benefited the elite at the expense of the poor.

To maintain a private common property regime it is vital that new land tenure regulations, which provide pastoralists with security of access to the key resources for the critical period, are promulgated (de Haan, 1996).

4.3 Implementation of policies promoting sustainable development

Desertification can only be successfully combated if poverty is alleviated and if an appropriate institutional framework for local development is developed. The example from Iceland (A. Arnalds, this volume) shows, however, that in industrialized countries actions must also be taken by the society. Advocated systems in DCs include a risk-minimizing component fulfilled through stabilization of producers income.

4.3.1 Actions for poverty alleviation

DC countries that rely heavily on rangelands for their livelihood are among the poorest in the world. Sixteen of the twenty countries in the Sudano-Sahelian region are among the most disadvantaged according to the United Nations Development Program Human Development Index (Soumare, 1996). The latter reported that in 1980 and 1987, only four of the 22 countries in this region achieved economic growth rates above 3% and eight registered a decline. In these countries, literacy rates are among the lowest in the world.

Severe environmental constraints such as shrinkage of grazing lands and chronic poverty have forced pastoralists to develop appropriate strategies of pastoral and agro-pastoral production in order to cope with recurring crises. However, the options available have been dramatically reduced over the last 30 years, increasing their vulnerability, especially to droughts. These constraints call for distinctive forms of assistance and development initiatives in order to allow poverty alleviation a *sine qua non* condition to the conservation and rehabilitation of the fragile rangeland and woodland resources.

4.3.2 Promotion of multiple use approach

There is a growing trend for development planners to look at rangeland as having multiple values, not just for animal production. Thus, instead of the narrow grazing perspective, recent strategies for rangeland management and planning insist on the critical importance of additional functions of rangeland including maximization of water harvesting, esthetics, biodiversity conservation and valorization. Such a vision makes rangeland rehabilitation programs more economically appealing as it contributes to improving marginal value of economic returns associated with restoration investments.

4.3.3 Adoption of opportunistic strategies and supporting actions

There is also a shift in paradigm towards a more realistic perception of the pastoral systems, their incentives and constraints. In this regard, the strategy of opportunistic management of rangelands is more attractive than the rigid conservative approach based on the carrying capacity model. Inspired by this philosophy, some DCs are currently devising development schemes, based on a system of use that takes advantage of the high production of forage in periods of above-average rainfall without degrading the lands in less favourable years. Such a system, which recognizes the resilience of the lands, requires for its fulfillment, in addition to strong local organization, supporting actions such as famine relief and emergency marketing strategies.

During droughts, undernutrition forces many animals to market, driving livestock prices down at a time when poor harvests are inflating grain prices. Under these adverse conditions, some pastoralists may starve unless assisted. A reasonable approach to relief provisioning would be to control speculation on food prices by securing adequate levels of grain availability in the market, at eventually subsidized prices. To manage this supply-demand balance, it would be necessary to rely on drought early-warning systems that provide information on the geographical extent and severity of the drought, and adequate transport infrastructure, to insure that food is moved in a timely manner and to the appropriate locations.

An important component of the strategy to mitigate drought impact is the establishment of improved livestock marketing systems. Rangeland inventories suggest that livestock population fluctuations are unavoidable when rainfall is erratic and variable. Under these conditions, a realistic goal is to design marketing systems which can absorb such fluctuations. Low cost techniques of meat preservation, access to the largest possible market for meat, and the elimination of subsidized international competition may be components of this effort.

5. CONCLUSIONS

Rangelands in DCs are presently at a turning point. Most assessments indicate that they are degraded and subject to heavy pressure and conflicting uses (see Arnalds and Archer, this volume). Trends in rangelands are country specific. Under current practices and prevailing socioeconomic conditions, the most likely scenarios are severe degradation or abandonment. Fortunately, a new line of thinking, emphasizing the concept of sustainable development is emerging worldwide, which may provide a brighter alternative to this grim picture. The lessons learned from the past also suggest that the development of rangelands depends on the establishment of appropriate institutional and policy frameworks (Sanders, this volume). When such a framework is in place, we can proceed with the implementation of biological improvements.

The suggested new approach to development problems is holistic in its philosophy, integrated in its planning, and co-ordinated in its implementation. This approach gives special attention to poverty alleviation and considers drought as an inherent phenomenon in the arid and semi-arid world, that needs to be dealt with as a structural problem. Drought management is therefore becoming an essential component of so-called opportunistic management, which integrates risk management during drought, by providing for mobility and flexibility in livestock numbers, and promoting other mechanisms including insurance arrangements, sales, increased slaughter and cold storage. This new approach is becoming possible because of the explosion of grassroots associations with growing lobbying power. In many DCs, a broad based civil society is operating in conjunction with public institutions.

Decentralization of decision-making on rangeland use is a recent, positive trend in rangeland development in some DCs. This process is, however, still at its very beginning and remains fragile. Thus, while the desirability of decentralization and participation is widely acknowledged, there are still some important constraints. Radical change in the way government agencies do business will be required. Participatory decision-making is an iterative process, radically different from the linear and deterministic ways projects have traditionally been planned.

Another positive development relates to the shift in perception of the value of local knowledge of ecological constraints and opportunities. The necessity to learn from local people is advocated everywhere. Current design of development plans includes more and more a decisive compromise between science and tradition, as it is doubtful that either one alone will produce the desired results (see also A. Arnalds, this volume; Sanders, this volume).

REFERENCES

Aidoud, A. 1994. Pâturage et désertification des steppes arides en Algérie: Cas de la steppe d'alfa. In: Proceedings of the Workshop on Desertification and Land Use in the Mediterranean Basin, 28–30 June 1993, Almeria, Spain. Parallelo No. 37, pp. 33–42.

Arnalds, A. and Archer, S. (eds.) 1999. Proceedings of the Rangeland Desertification, International Workshop, September 1997, Iceland. RALA Report No. 200. (In press).

Behnke, R.H. and Scoones, I. 1992. Rethinking range ecology implications for rangelands management in Africa. Environment Working Paper No. 53. World Bank, June 1992.

Bilsborrow, R.E. and Okoth-Ogendo, H.W.O. 1992. Population-driven changes in land use in developing countries. Ambio 21(1), 37–45.

Biot, Y., Lamber-t, R. and Perkin, S. 1992. What's the problem? – An essay on land degradation, science and development in Sub-Saharan Africa. School of Development Studies, Discussion Paper No. 22. February 1992.

Chisholm, A.H. 1992. Australian agriculture: a sustainability story. Australian Journal of Agricultural Economics 36(1), 1–29.

Daily, G.C. 1995. Restoring value of the world's degraded lands. Science 269, 350–354.

de Haan, C. 1996. Rangelands in the developing world. In: Proceedings of the fifth International Rangeland Congress, Salt Lake City, Utah, USA, July 23–28, 1995. Ed. N.West. pp. 180–184.

Holdgate, M.W. 1994. Ecology, development and policy. Lecture delivered at the University of Stirling, Scotland, on January 5, 1994.

Hubert, B. 1991. Changing land uses in Provence (France). Multiple use as a management tool. Options Méditerranéennes No. 15, pp. 31–52.

Kallala, A. 1994. La lutte contre la désertification en Tunisie: Bilan et perspective. In: Proceedings of the Workshop on Desertification and Land Use in the Mediterranean Basin, 28–30 June 1993, Almeria, Spain. Parallelo No. 37, pp. 205–210.

Kates, R.W., Hyden, G. and Turner, B.L. 1993. Theory, evidence, study design in population growth and agricultural change in Africa. In: Population Growth and Agricultural Change in Africa. Eds. B.L. Turner, G. Hyden and R.W. Kates. University of Florida Press, Gainsville, Florida.

MAMVA. 1994. Etude de la consommation nationale de bois de feu au Maroc. Direction des Eaux et Forêts et de la Conservation des Sols, Rabat, Morocco.

MAMVA. 1995. Stratégie de développement des terres de parcours au Maroc. Schéma Directeur d'Aménagement des Parcours. Etude coordonnée par H. Narjisse et réalisée par M. Amane, 0. Berkat, H. Narjisse et M. Tozy. Direction de l'Elevage, MAMVA, Rabat, Morocco.

Mortimore, M. 1992. Environmental change and drylands management in Machakos district, Kenya, 1930–90. ODI Working Paper No. 63, May 1992.

Narjisse, H. 1996a. Plan d'Action National pour l'Environnement: Sol et Environnement. Ministère de l'Environnement, Rabat, Morocco.

Narjisse, H. 1996b. The range/livestock industry in developing countries: current assessment and prospects. In: Rangelands in a Sustainable Biosphere. Ed. N. West. Proccedings of the Fifth International Rangeland Congress, Salt Lake City, Utah, USA, July 23–28, 1995.

Naveh, Z. 1994. A holistic approach to landscape degradation and desertification. In: Proceedings of the Workshop on Desertification and Land Use in the Mediterranean Basin, 28–30 June 1993, Almeria, Spain. Parallelo No. 37, pp. 97–105.

Nessim, A. 1996. Economic, Social and Cultural Causes and Consequences of Drought and Desertification, Including Linkages to Poverty, Population Pressure, Food Security, International Trading Patterns, Traditional Mechanisms for Coping with Drought and Desertification, and Gender/Religion Aspects. IFAD, Technical Advisory Division, Rome, Italy.

Niamir-Fuller, M. 1996. A new paradigm in African range management policy and practice. In: Proceedings of the Fifth International Rangeland Congress, Salt Lake City, Utah, USA, July 23–28, 1995. Ed. N. West.

Olsson, L. 1993. On the causes of famine-drought, desertification and food security in the Suddan. Ambio 6.

Soumare, M. 1996. Role of Planning Systems and Instruments, Including Integration of Antidesertification Programs into Overall Development Programmes. UNSO.

UNCED 1992. Agenda 21. United Nations, New York.

UNEP 1996. Status of Desertification and Implementation of the United Nations Plan of Action to Combat Desertification. UNEP-GRID, Sioux Falls.

The United Nations Convention to Combat Desertification: constraints to implementation in Eastern Africa

Naftali Manddy Onchere
EcoNews Africa, P.O. Box 76406, Nairobi, Kenya
Tel: 254 2 721076/99 or 721655; Fax: 254 2 725171; E-mail: <onchere@iconnect.co.ke>

ABSTRACT The United Nations Convention to Combat Desertification (CCD) and mitigate the effects of drought and desertification was adopted in June 1994. Despite the United Nations Resolution for Urgent Action for Africa, the implementation of the CCD in the rangelands of Eastern Africa has been constrained by the often intricate patterns of land use and cropping exacerbated by drought, civil strife and famine (Anon., 1997; Guturo, 1997; Njuki, 1997). There have been a number of attempts to implement the CCD at subregional, national and local level. Experiences gained from these attempts have identified major constraints specific to each level. Overcoming these constraints will improve implementation of the CCD and reverse or slow the degradation of Eastern African rangelands. At the subregional level, the main constraints to the implementation of the CCD were related to poor communication and information brokerage (Anon., 1995), inadequate technical and human resource capacity, lack of political goodwill and low commitment of subregional intergovernmental organisations; ineffective famine early warning and drought monitoring (Henricksen and Durkin, 1986; Cutler 1985); insufficient capacity to monitor and control pests and pestilence (Gathuru et al., 1991; Latigo et al., 1989; Onchere and Scott, 1989; Onchere and Odiyo, 1993); and scanty or faulty donor support for subregional and cross border initiatives. At the national level, the main constraints to the implementation of the CCD were related to inadequate or draconian legislation (Abdalla, 1993); an unfavourable economic and funding climate; low priority accorded to environmental and pastoral-related issues (Rutten, 1992; Dransfield, 1994); debilitating civil strife; sporadic famine; inadequate technical and personnel capacity; lack of political goodwill and low commitment of national consultative structures and mechanisms; poor infrastructure; and inappropriate or unclear donor policies (Guturo, 1997). At the local level, the constraints to the implementation of the CCD include drought; civil strife and livestock rustling; famine; poverty; unfavorable economic climate and legislation; cultural and gender barriers; poor technical and human resource capacity of local populations; inappropriate technologies; and top-down government policies.

Key words: CCD, desertification, drought, IGAD, SADC, SRAPs.

1. INTRODUCTION

As a follow-up to the recommendations of Chapter 12 of Agenda 21 from the UNCED "Earth Summit" Conference in 1992, the UN General Assembly has elaborated the Convention to Combat Desertification (CCD) in countries experiencing serious drought and/or desertification, particularly in Africa (Anon., 1997). The CCD was adopted in June 1994 and entered into force in December 1996. The CCD sets out a framework within which governments should prepare action programmes aimed at combating desertification and mitigating the effects of drought at national, subregional, regional and global levels. At the local level, communities are also asked to elaborate Local Area Development Programmes (LADPs), as a basis for NAPs (National Action Programmes), to combat desertification and mitigate the effects of drought. Subregional Action Programmes (SRAPs) and the Regional Action Programme (RAP) for Africa will also be elaborated according to the Implementation Annex for Africa contained in the CCD (United Nations, 1994).

The CCD promises to drastically reshape the international aid process by engaging donor nations and agencies in a new partnership developed through a consultative process. The aim is to ensure that funding of programmes are better coordinated, that funding is based on the needs of the affected countries, that donors can be sure their funds are well-spent, and that recipients obtain the maximum benefit. The CCD emphasizes a bottom-up approach with strong local participation in decision-making. It puts local communities on equal footing with other actors in the development process. Communities and their leaders, as well as Non-Governmental Organisations (NGOs), experts and government officials will work closely together to formulate action programmes. The CCD recognizes that desertification is primarily a problem of sustainable development, poverty and human well-being, and environmental preservation. Social and economic issues, including food security, migration and political instability, are closely linked to land degradation. So are such issues as climate change, biological diversity, and fresh water supplies (Anon., 1997). The CCD emphasizes the need to coordinate research efforts and action programmes for combating desertification with those related to these concerns.

The Eastern African rangelands comprise large portions of Tanzania, Kenya, Uganda, Sudan, Eritrea, Ethiopia, Djibouti and what was once Somalia (now Somalia and Somaliland). These countries are members of the IGAD (Inter-Governmental Authority on Development) subregion, except for Tanzania, which is a member of the SADC (Southern African Development Cooperation) subregion. Within the Eastern African rangelands and under the special proviso of the UN Resolution for Urgent Action for Africa, the implementation of the CCD was to start immediately after the adoption of the CCD in June 1994 and before its formal ratification in 1996. Sudan was the first to ratify the CCD (24 November 1995) followed by the newly independent Eritrea (14 August 1996). The other countries (Djibouti, Tanzania, Kenya, Uganda and Ethiopia) ratified the CCD in June 1997, so as to be on time to fully participate in the first Conference of the Parties (COP 1) to the CCD (CCD Secretariat, 1997). Civil strife and inadequate international recognition of the several warring factions have complicated the case of Somalia.

2. IMPLEMENTATION AT SUBREGIONAL LEVEL

In the past, efforts to combat desertification at the sub-regional level were done in a sectoral manner. Governments, NGOs, donors and community groups seldom worked together. By contrast, the preparation and implementation of the sub-regional action programmes (SRAPs) call for co-operation and partnership among the principal actors (IGAD, 1997; SADC, 1997).

IGAD and SADC were designing their SRAPs in August 1997, through consultations among the affected countries of each subregion. These SRAPs are meant to better promote co-operation among scientific and technical institutions, improve cross-border transport and communication, and promote joint programmes for the sustainable management of shared rivers and other cross-boundary ecosystems.

SRAPs require that institutions within the subregion, with their diverse capacities, capabilities and scope of operation, work in a clearly coordinated manner to achieve the goals of the SRAP. The elaboration of the SRAPs involved the participation of governmental and NGOs, donor agencies and international and United Nations agencies. NGOs represented included EcoNews Africa, the International Union for the Conservation of Nature (IUCN), Observatoire du Sahara et du Sahel (OSS), the National NGO Coordinating Committee of Kenya, Pastoralist Environmental Network for the Horn of Africa (PENHA) from Ethiopia, Sudanese Society for Environmental Conservation, Uganda Women Tree Planting Movement, Yonge Nawe of Botswana, Commutech and Zero of Zimbabwe and Journalists Environmental Association of Tanzania (JET).

In the IGAD subregion, the SRAP supporting bureau will be the implementing body. The IGAD SRAP Steering Committee will oversee its tasks and activities. The support bureau will also act as the Secretariat of the Steering Committee; review and coordinate the SRAP implementation; liaise with the Convention's Secretariat and the African Regional Coordinating Unit (RCU) for desertification control; and promote any other measures necessary to increase the efficiency of the NAPs and SRAP (IGAD, 1997).

3. IMPLEMENTATION AT NATIONAL LEVEL

A number of countries within the rangeland areas of Eastern Africa were elaborating their NAPs in August 1997. A few countries had also started elaborating and designing National Desertification Funds (NDFs). Kenya and Uganda held their first National Consultative Forums to begin articulating NAPs in October and December 1998, respectively. In Uganda, Kenya and Tanzania, cabinet papers have been prepared for the establishment of National Desertification Funds (NDFs); these now await ratification. Tanzania, Ethiopia, Eritrea, Sudan, Djibouti, Somalia and Somaliland are planning to hold National Consultative Forums to discuss NAPs in summer of 1999. Many of the NAPs elaborated have not been methodologically in the spirit of the CCD. They have tended to be top-down and have not involved local populations. A number of NGOs have been invited to join the NAP processes. Participation of civil society, in particular the affected populations, at this level as well as in the elaboration of the NDFs appears to have been hurried and not well-planned in any of the countries.

Pastoralism is practised across borders in many African countries without any regard for political boundaries. In many cases, the pastoralists living across a border are from the same ethnic group, and even the same family, but may be called by different names in the neighbouring countries (Anon., 1997; Dransfield, 1994; Haro, 1997; Njenga, 1997). These pastoralists, who live in the rangeland areas affected by drought and desertification, have generally not been consulted or involved in the elaboration of NAPs and NDFs. In Kenya and Uganda for instance, elaboration of the NAPs and NDFs, have not included representation of the Maasai, Kalenjin and Karamojong pastoralists nor the Acholi and Langi tribesmen, who have a large stake in efforts to combat desertification and cope with drought.

4. IMPLEMENTATION AT LOCAL COMMUNITY LEVEL

Awareness of the CCD is very low at the local community level. As a result, implementation at the local community level has been haphazard and uncoordinated (Anon., 1997). Available information suggests that only the Olkernerei Indigenous Pastoralist Survival Programme of Munduli District of northern Tanzania, the Miji Kenda women groups at Gede in Malindi district and the Samburu and Rendile pastoralists of Northern Kenya have developed Local Area Development Programmes (LAPDs) as envisaged by the CCD. Unfortunately, these LAPDs have not had impact on any of their respective country's NAPs or NDFs.

5. CONSTRAINTS TO CCD IMPLEMENTATION AT SUBREGIONAL LEVEL

5.1 Political strife

During early 1997, there was an outcry from communities in the northern regions of Kenya and Uganda and the southern regions of Ethiopia and Sudan, regarding the high incidences of livestock rustling, several of which culminated in massacres. Insecurity and conflicts over the use of natural resources are a major constraint to the implementation of the CCD within IGAD. Although livestock rustling is not new in the region, it is obvious that inter-governmental agreements may not resolve the problem unless the local populations are involved in negotiations. The Kenyan and Ethiopian governments have been trying to enforce peace without involving the affected local populations, with little success. Attempts by civil society and local community-based organisations to broker peace between the Turkana and Karamojong pastoralists of Kenya and Uganda, respectively in early 1997, were seen as usurping the power of the elected governments of the day. These attempts were crushed using draconian laws of sedition and maintenance of public security act. IGAD has never seen it fit to get involved in such cross-border issues. This can only be attributed to inadequate capacity and political goodwill within IGAD compounded by poor communication and information brokerage within these remote and underdeveloped rangelands.

5.2 Pestilence and famine

Inadequate control of crop pests also constrains CCD implementation. There are many species of migratory pests that ravage food crops and pasture land. Notable among these are the desert locust (*Schistocerca gregaria*) of biblical fame, *quelea quelea* and the African armyworm (*Spodoptera exempta*). The comprehensive monitoring, survey and control of these pests continues to elude national plant protection departments: the Desert Locust Control Organization for Eastern Africa (DLCO-EA) (Gathuru et al., 1991; Latigo et al., 1989; Onchere and Scott, 1989; Onchere and Odiyo, 1993); the Regional Centre for Services in Surveying, Mapping and Remote Sensing (RCSSMRS); and the International Centre for Insect Physiology and Entomology (ICIPE). Equipment for the control of these pests continues to be either inappropriate or beyond the reach of most pastoralists (Amisi et al., 1991; Latigo et al., 1989). Strategies to replant crops where migratory pest damage has been severe are not always economically rational, depending on the stage of the crops when infested, whether control was undertaken or not, the amount and duration of rainfall after infestation and subsequent cultural practices (Gathuru et al., 1991; Onchere and Scott, 1989; Onchere and Odiyo, 1993).

Throughout the Eastern African rangelands, famine early warning and drought monitoring are ineffective or inadequate at all levels from the subregional to the local level (Ayling, 1993; Curran, 1983; Cutler, 1985). The famine early warning project at the RCSSMRS had very little impact on the food policy and agricultural extension services within Eastern Africa. Despite clearly forecast poor crop yields in western Kenya and southern Sudan in 1994, the departments of agriculture and other development agents concerned did not react in time to avert the serious food shortages that followed (Henricksen, personal communication). Although the experience gained in the operation of satellite-based early warning systems suggests that stratification of the African region into coherent geographical areas would improve forecasting over traditional ground data sources, support to such continues to be lukewarm, and perhaps even declining (Henricksen, 1986; Cutler, 1985).

5.3 Unresolved issues

The draft SRAPs for IGAD (February 1997) and for SADC (July 1997), it is noted, did not enumerate priority activities, nor identify the institution or sector responsible or where the resources needed for implementation would come from. It is also felt that there is an urgent need to provide direction to raise awareness and to develop innovative fund-raising strategies for implementing the SRAP.

A number of issues remain unresolved. How could the SRAP forums start elaborating the IGAD and SADC SRAPs when not one NAP had been formulated? To what extent was the process of elaborating the SRAP community-driven? How different were the planning processes for the forums vis-à-vis the usual top-down intergovernmental planning and implementation processes? Were the forums a way of putting new wine in old wineskins or putting old wine in new wineskins? How committed are the governments of Eastern Africa in formulating the SRAPs when, with the exception of Sudan and Eritrea, they have apparently only ratified the Convention so as to participate in COP 1?

A number of NGOs were invited to the SRAP forums, some being included in the government delegations. The participation of affected populations, at this level as well as in the implementation of the SRAPs, appear not to have been well planned. Participation of NGOs in the implementation of the IGAD SRAP occurred only after intense lobbying of government delegates. This was the case notwithstanding the fact that it was generally acknowledged that to sustain some of the priority programmes, it could be vital for local communities to assume ownership of the programmes (IGAD, 1997). Whether an acknowledgement of the importance of involving affected local populations will translate into their involvement remains to be seen.

The forums recognized that in addition to the traditional financial mechanisms and arrangements, there is an urgent need for new and flexible funding mechanisms. For the IGAD SRAP, it was agreed that facilitation activities would include raising awareness, preparation of sensitization materials, training of task force members in the nine priority areas of the IGAD SRAP, and institutional networking. Supporting the IGAD CST (Committee on Science and Technology) and the sub-regional steering committee, project preparation and the assessment of the state of desertification were other agreed activities. A time-frame for each of the activities could not be agreed upon. It was recommended that the IGAD Secretariat implement the SRAP according to its current work programme and available resources (IGAD, 1997).

In the Draft IGAD SRAP of February 1997, a steering committee and task force were considered necessary in the implementation of SRAP. The IGAD SRAP Steering Committee included the NAP coordinators within the national CCD focal points. The Steering Committee will provide policy guidance, review the progress of the SRAP, facilitate the mobilization of resources and ensure harmonization of the projects developed under NAPs and the SRAP. The Steering Committee will also establish and ensure data collection and dissemination and develop linkages with NGOs and relevant regional and Sub-regional organisations involved in the implementation of the Convention (IGAD, 1997).

6. CONSTRAINTS TO CCD IMPLEMENTATION AT NATIONAL LEVEL

At the national level, a number of factors constrain the effective implementation of the CCD. An enabling legal environment is generally lacking and there are numerous draconian laws, most dating back to pre-independence periods (Guturo, 1997; Mwangi, 1997; Njenga, 1997; Njuki, 1997). These laws criminalize fund-raising for local projects, and restrict freedom of speech, assembly, expression and association. They do not allow for local community consultative structures but do allow for the plunder of local community resources by government officials and well-connected foreign operatives. In some instances, the funds and activities of NGOs and Community-Based Organisations (CBOs) are directly determined and restricted (not just regulated) by the government which must be represented in these local organisations (Onchere and Guturo, 1997).

The rangeland areas of Eastern Africa have been traditionally neglected and financially exploited by governments. These areas contribute a large proportion of the national economies. Environmental and pastoralist related issues on rangelands are accorded low priority (Dransfield et al., 1994). This has led to poor infrastructure and unfavorable economic and funding climate for the indigenous peoples of the Eastern African rangelands. These circumstances, combined with the fragile nature of living conditions in the rangelands, the uncertain nature of crop and livestock yields and the struggle for resources to sustain livelihoods, have invariably led to civil strife and sporadic famine. Generally, the affected governments have inadequate technical and personnel capacity to combat desertification or to mitigate the effects of drought and desertification (Anon., 1997). These factors include a brain drain from the public sector to the corporate, private and civil sectors within the affected countries or to the more economically rewarding economies of Southern Africa, Europe or North America.

Local pastoral communities in the Eastern African rangelands play a very small and docile role in national politics as compared to those communities from cash crop areas (Dransfield et al., 1994). Given their small collective voice, most governments therefore have minimal political goodwill and commitment to rural communities from the rangeland areas. The plight of these communities has been further exacerbated by inappropriate and unclear donor policies that have been used to promote inappropriate and deleterious technologies (Abdalla, 1993; Anon., 1997; Dransfield et al., 1994; Onchere and Nyang, 1996; Onchere and Musyoka, 1994; Onchere et al., 1993; Onchere et al., 1992; Rutten, 1992; Sanders, this volume; Narjisse, this volume).

There have been a number of institutional constraints to CCD implementation at national levels. These have included the fact that the designated National Coordinating Bodies (NCBs) lack authority or credibility to effectively bring relevant departments together. The NCBs rarely include NGO or CBO representatives. The Coordinator of the NCB is often overburdened, generally working under very tight schedules and lacking adequate support (Anon., 1997).

Procedural and methodological constraints are also problematic. Often, there has been neither a vision nor a clear road map for the process. There has been a lack of understanding that the value of a NAP lies not in the final document, but rather in the process of participatory planning and policy-making. It has proved to be extremely difficult to operationalize a participatory process. Although the CCD calls for integration, several NAP consultations have been run in parallel, creating a risk of "consultation fatigue".

Funding and coordination constraints have also generally hampered the smooth implementation of the CCD at national level. Lack of awareness, and thus commitment, by the Finance\Planning and Foreign Ministries has been a major constraint. Other factors have included difficulties in harmonizing external inputs in the early planning process and getting donors to participate in the national consultative forums. Continuous support required by a process-oriented approach poses problems to many donor procedures.

7. CONSTRAINTS TO CCD IMPLEMENTATION AT LOCAL COMMUNITY LEVEL

At the local community level, the smooth implementation of the CCD has been constrained by top-down national policies which place the bulk of development activities on national governments that are already overstretched and hard-pressed to provide basic services of security, infrastructure, health and education. These government policies usually assume a monopoly of knowledge and are frequently extending inappropriate and non-sustainable technologies (Onchere and Musyoka, 1994; Sanders, this volume; Narjisse, this volume). High external input agriculture is normally propagated as the way to alleviate poverty and improve food production, notwithstanding the fact that these technologies are neither affordable to the already impoverished local populations nor sustainable (Dransfield et al., 1994; Onchere and Musyoka, 1994). In the rangelands, literacy levels are invariably lower than in cash crop areas in Eastern Africa. The capacity of these local populations to combat desertification and mitigate the effects of drought and desertification is at the lowest in living memory (Rutten, 1992). In Kenya the national per capita income stands at US $280, life expectancy at 58 years and infant mortality rates at 56 per thousand births. In the Kenyan rangelands however, child mortality rates stand at 156 per thousand and 30–50% of the population have no guarantee for household food security even under normal and favorable weather conditions. The districts of Marsabit, Samburu, Isiolo and Turkana, for example, have well over 70% of their populations living below the poverty line (SNV-Kenya, 1996).

At the local environmental management level in Kenya, for instance, it is difficult to control migrant livestock herds during drought or cattle rustling; the use of areas set aside as pasture reserves; and charcoal burning activities around trading centres (Haro, 1997). Negotiations with intruders from other areas who claim user-rights on limited pastures in accordance with age-old customs and traditions can not be controlled. An alien policy environment that lacks a locally clear policy on gender and development makes it difficult for socio-culturally gender sensitive initiatives to be addressed (Anon., 1993; Horowitz and Jowkar, 1993; Njenga, 1997; Sanders, this volume). Traditions, customs and the overall constraints of a patriarchal system determine the position of women in the family and the community within the rangelands of Eastern Africa. The Maasai women of Kenya and Tanzania for instance are responsible for house building. The effects of drought and desertification have decreased the availability and access to building materials to

Maasai women. The plight of these women, and other women within the Eastern African rangelands in general, has been worsened by the fact that they are rarely represented in village-level deliberations. These women have no rights to land or property and have very low levels of education (Njenga, 1997).

Availability of extension staff is inadequate. Agents generally visit villages or farm households a maximum of twice a year, leaving farmer-to-farmer interactions as the most sustainable technology transfer mode (Onchere and Nyang, 1996; Onchere and Wituka, 1995).

8. CONCLUSIONS AND RECOMMENDATIONS

The 1997 draft SRAPs for IGAD and SADC have been hurridly elaborated with inadequate input from local communities who are the landowners and resource users. These drafts have neither designated nor identified organisations that will ensure the implementation of the SRAPs. They have not taken stock of the major gaps and constraints that will hinder SRAP implementation. They have not addressed the inadequate capacities of the affected local populations and sub-regional institutions nor their priorities. The SRAPs should therefore include:

- Designation of sub-regional organs and institutions to oversee SRAP development.
- Sensitization, awareness-raising and training of relevant personnel working in the collaborating institutions identified.
- Identification of the gaps and constraints, as well as measures to enable the smooth implementation of the SRAP process.
- Structuring and strengthening of appropriate institutional arrangements.
- Carrying out assessments within the subregion and recording the progress made.
- Identification of financial mechanisms and ways to mobilize financial resources.

The subregional secretariat should act as an overall coordinator, facilitator and catalyst to the SRAP. However, specific institutions and member states operating as one or as a network should do the formulation and implementation of SRAP. The institutions identified for any SRAP programme should designate a networking organ with linkages and a focal point. The subregional secretariats should facilitate the design, establishment and sustainability of networks and information systems through formal arrangements, contractual agreements or memoranda of understanding.

At the national level, the NAP process should be seen in two phases: a preparatory phase that leads to a national forum which marks the beginning of the implementation phase; and an implementation phase that follows genuinely elaborated NAPs.

At the local community level, LAPDs should be elaborated so as to guide the nature and manner of implementation of the NAPs as envisaged in the implementation of the CCD at the national level.

ACKNOWLEDGEMENTS

I wish to acknowledge the willing cooperation of the Netherlands Development Organization as well as EcoNews Africa during the preparation of this paper, especially with respect to providing much of the funding and opportunity necessary for information gathering. The organizing and supporting committee of the Icelandic Rangeland Desertification Workshop, especially Olafur Arnalds, Ulfur Bjornsson and Steinunn Geirsdottir are also gratefully acknowledged.

The loving support of my wife Edith and children Martin, Peris and Peninah made all the information gathering worthwhile and rewarding.

REFERENCES

Abdalla, A.J. 1993. Pastoral land rights. Paper presented at the International Workshop on Listening to the People: Social Aspects of Dryland Management at UNEP, Nairobi. UNDP/OPS, Unpublished typescript.

Amisi, J.L., Bahana J., Mallya, G., Konchellah, D.L., Onchere, N.M. and Dewhurst, C.F. 1991. A summary report of aerial surveys and spray trials with a Bell Jet Ranger helicopter against the African armyworm, *Spodoptera exempta* (walk.) (Lepidoptera, Noctuidae), in the primary outbreak areas of central Tanzania during the December 1989 – January 1990 season. DLCO-EA Technical Report No. 101.

Anonymous 1993. The importance of participatory approaches for dryland management and anti-desertification programmes. Based on case studies from Burkina Faso, Ghana, Kenya and Zimbabwe. INCD, Unpublished typescript.

Anonymous 1995. The convention on desertification and the role of the media. Paper prepared for the IGAD NGO consultative forum for the subregional action programme in Asmara, Ethiopia. Unpublished typescript.

Anonymous 1997. Implementing CCD at national level. Update August 1997. UNSO\UNDP.

Ayling, R.D. 1993. Departure points: Researchers and Rural Communities. Paper presented at the International Workshop on Listening to the People: Social Aspects of Dryland Management at UNEP, Nairobi. IDRC, Unpublished typescript.

CCD Secretariat. 1997. Status of entry into force of the CCD. August 1997.

Curran. P.J. 1983. Multispectral remote sensing for the estimation of green leaf area index. Phil. Trans. R. Soc. 309A, 275.

Cutler, P. 1985. Review of progress made towards instituting technical and institutional improvements in the early warning system. Report prepared by the Relief and Rehabilitation Commission and UNICEF, Addis Ababa.

Dransfield, R., Brightwell, R., Kanunka , J., Kamanga, J. and Adan, I. 1994. Strengthening resource management and the pastoralist economic system for sustainable development of the Kenyan rangelands, SNV Kenya.

Gathuru, P.N., Onchere, N.M. and Scott P.J. 1991 The assessment of crop loss attributable to the African armyworm, *Spodoptera exempta*, (walk.) (Lepidoptera, Noctuidae), using a paired plot technique. A case study. DLCO-EA Technical Report No. 100.

Guturu, F.S. 1997. The Role of NGOs in combatting desertification in Kenya. Paper presented at the National Workshop on Desertification, Nairobi. August 1997. Unpublished typescript.

Haro, G.O. 1997. Participatory approaches in community consultations and implementation of anti-desertification activities; experiences from South-Western Marsabit, Northern Kenya. Paper presented at the National Workshop on Desertification, Nairobi. August 1997. Unpublished typescript.

Henricksen, B.L. and Durkin, J.W. 1986. Growing period and drought early warning in Africa using satellite data. Int. J. Remote Sensing 7, 1583–1608.

Horowitz, M.M. and Jowkar, F. 1993. Gender and participation in the conception, planning, and implementation of environment and development projects in the drylands. Background paper presented at the International Workshop on Listening to the People: Social Aspects of Dryland Management at UNEP, Nairobi. Institute for Development Anthropology, Unpublished typescript.

IGAD 1997. Draft IGAD Sub-regional action programme to combat desertification. February 1997.

Latigo, A.A.R., Dewhurst C.F., Wambugu, N. and Onchere, N.M. 1989. The evaluation of the migrant pest sprayer (STU) for armyworm control in Kenya. DLCO-EA Technical Report No. 97.

Mwangi, W. 1997. Bridging the gap between policy makers and communities in combatting desertification. Paper presented at the National Workshop on Desertification, Nairobi. August 1997. Unpublished typescript.

Njenga, C.K. 1997. Implementing the CCD – What role for women? Paper presented at the National Workshop on Desertification, Nairobi. August 1997. Unpublished typescript.

Njuki, V.K. 1997. Review of CCD implementation in Kenya. Paper presented at the National Workshop on Desertification, Nairobi. August 1997. Unpublished typescript.

Onchere, N.M. and Guturo, S. 1997. A guide kit to assist local communities to implement the United Nations Convention to Combat desertification and mitigate the effects of drought. EcoNews Africa, National NGO Coordinating Committee on Desertification (NCCD), and Environmental Liaison Centre International (ELCI). August 1997. (In press).

Onchere, N.M. and Nyang, M.N. 1996. First impact monitoring report with specific reference to agroforestry activities in Uasin Gishu. Kenya Woodfuel and Agroforestry Programme (KWAP) technical report, March 1996.

Onchere, N.M. and Wituka C.G. 1995. The experience of the Kenya Woodfuel and Agroforestry Programme (KWAP) Uasin Gishu, in agroforestry extension and topics for research. Paper presented at the ODA/KARI (Overseas Development Administration/Kenya Agricultural Research Institute) organised planning workshop at Kitale.

Onchere, N.M. and Musyoka, J.A.M. 1994. New developments in farmer participatory approaches: The Case of Kenya Woodfuel and Agroforestry Programme's (KWAP) agroforestry farming systems approach (AFSA). KARI/ISNAR Workshop at Nyeri. February 1994.

Onchere, N.M., Nyang, M.N., Musyoka, J.M., Jami, A.M., Kola, H.O., Wanjau, S.M., Moenga, J.M., Onyango, S.M., Wambugu, C.M. and Onamu, R. 1993. Participatory monitor-

ing and evaluation. Manual for training subject matter specialists (SMS). Kenya Woodfuel and Agroforestry Programme (KWAP); ETC Kenya Consultants BV.
Onchere, N.M., Nyang, M.N., Musyoka, J.M., Jami, A.M., Kola, H.O., Wanjau, S.M., Moenga, J.M., Onyango, S.M., Wambugu C.M. and Onamu, R. 1993. Participatory monitoring and evaluation. Manual for training frontline extension staff (FLS). Kenya Woodfuel and Agroforestry Programme (KWAP); ETC Kenya Consultants BV.
Onchere, N.M. and Odiyo, P.O. 1993. The distribution, onset and spread of the African armyworm, *Spodoptera exempta*, (walk.) (Lepidoptera, Noctuidae), outbreaks in Eastern Africa for the period 1 October 1990 – 30 September 1991. GIS'93 Symposium; Vancouver, British Columbia, Canada.
Onchere, N.M., Nyarangi, E.G. and Howorth, C.N. 1992. Woody biomass market survey in Kisii and Kakamega districts. Kenya Woodfuel and Agroforestry Programme (KWAP); ETC Kenya Consultants BV.
Onchere, N.M. and Scott, P.J. 1989. The assessment of the crop loss attributable to the African armyworm, *Spodoptera exempta*, (walk.) (Lepidoptera, Noctuidae), and of a strategy of replanting where damage is severe. A case study. DLCO-EA Technical Report No. 98.
Rutten, M. 1992. Selling Wealth to Buy Poverty. The Process in Individualization of Land Ownership among the Maasai Pastoralists of Kajiado District, Kenya, 1890–1990. Nijmegen Studies in Development and Cultural Change. Vol. 10. Verlag Breitenbach Publishers, Saarbrucken.
Sabine, H. 1993. Listening to the people – Some reflections on the use of indigenous knowledge in strategies to curb environmental degradation. Paper presented at the International Workshop on Listening to the People: Social Aspects of Dryland Management at UNEP, Nairobi. Institute of Social Research.
SADC 1997. SADC Sub-regional action programme to combat desertification in Southern Africa. August 1997.
SNV-Kenya 1997. SNV-Kenya Policy plan, 1997–2002. SNV, Kenya.
UNITED NATIONS 1994. United Nations Convention to Combat Desertification in those countries experiencing serious drought and/or desertification, particularly in Africa. UNEP, Nairobi, Kenya. Printed 1995 in Switzerland.

MIX
Papier aus verantwortungsvollen Quellen
Paper from responsible sources
FSC® C105338

If you have any concerns about our products,
you can contact us on
ProductSafety@springernature.com

In case Publisher is established outside the EU,
the EU authorized representative is:
Springer Nature Customer Service Center GmbH
Europaplatz 3, 69115 Heidelberg, Germany

Printed by Libri Plureos GmbH
in Hamburg, Germany